# Lecture Notes in Mathematics

Edited by A. Dold and B. Eckmann

1157



Harold Levine

# Classifying Immersions into $\mathbb{R}^4$ over Stable Maps of 3-Manifolds into $\mathbb{R}^2$

Springer-Verlag
Berlin Heidelberg New York Tokyo

**Author**

Harold Levine
Department of Mathematics, Brandeis University
Waltham, MA 02254, USA

Mathematics Subject Classification (1980): 57R42, 57R45

ISBN 3-540-15995-9 Springer-Verlag Berlin Heidelberg New York Tokyo
ISBN 0-387-15995-9 Springer-Verlag New York Heidelberg Berlin Tokyo



Printing and binding: Beltz Offsetdruck, Hemsbach/Bergstr.
2146/3140-543210

TABLE OF CONTENTS

ABSTRACT

Let $f$ be a stable map from a compact, oriented, three manifold $M$ into $\mathbb{R}^2$. The main object of these notes, treated in Chapter III, is the classification of immersions $(f,h)$ of $M$ into $\mathbb{R}^2 \times \mathbb{R}^2$. (Here, immersions $(f,h_0)$ and $(f,h_1)$ are equivalent if they are connected by a regular homotopy, $(f,h_t)$.) A descriptive device used for this study is the space $W_f$ obtained by identifying points of $M$ that belong to the same component of the $f$-fibre. In Chapter I, the local geometry of $W_f$ is studied and all local descriptions of $W_f$ are given. In Chapter II, semi-local canonical coordinate expressions for $f$ near the singular set of $f$ are derived.

Partially supported by NSF

## Introduction

Our primary objective in this work is to study stable maps of compact three-dimensional manifolds into the plane. Stable maps generally are those whose character is unchanged by small perturbations, i.e. any small perturbation of a stable map can be obtained from it by composition with diffeomorphisms of the source and target manifolds. These maps have been characterized by John Mather [Mather, $G^2$].

Let $f$ be a smooth map of a compact three-dimensional manifolds, $M$, into the plane. The regular points of $f$ are those at which the Jacobian of $f$ has rank 2 and the singular points, $S(f)$, are the non-regular points. The stability of $f$ is equivalent to the following:

At each $P \in S(f)$, there are coordinates $(u,x,y)$ centered at $P$ and $(V,X)$ centered at $f(P)$ such that $(V(f(u,x,y)),X(f(u,x,y)))$ has one of the following expressions:

$$\begin{cases} (u,x^2+y^2) & , \quad P \text{ a definite fold point} \\ (u,x^2-y^2) & , \quad P \text{ an indefinite fold point} \\ (u,y^2+ux-\frac{x^3}{3}), & \quad P \text{ a cusp point.} \end{cases}$$

In addition no cusp point is a double point of $f \mid S(f)$ and on $S(f)-\{\text{cusps}\}$, $f$ is an immersion with normal crossings.

In attempting to understand the map $f$, we consider first the possible fibres of the map. Obviously, above every point in $\mathbb{R}^2 - f(S(f))$, the $f$-fibre is a set of disjoint embedded circles. However for $P \in S(f)$, the part of the $f$-fibre over $f(P)$ in the three above mentioned neighborhoods are an isolated point, two intersecting line segments, and a cusp (i.e. $\{u = 0,\ y^2 = x^3/3\}$). Thus the connected component of the $f$-fibre through a definite fold point is just the point itself and through a cusp point is:

However, on the connected component of the $f$-fibre through an indefinite point there may be one or two indefinite points. Thus those connected components of the $f$-fibres are connected graphs with one or two 'X-nodes'. The distinct possibilities are:

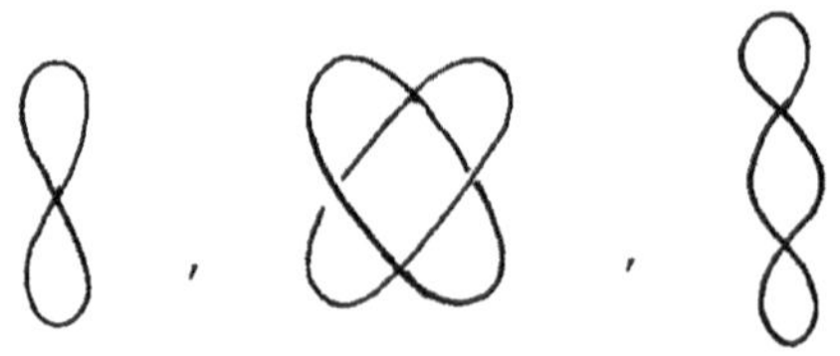

if M is orientable, and in addition if M is not orientable:

An indefinite point is called simple if it is the only point of $S(f)$ on the $f$-fibre component through it.

Call a subset of $M$, saturated if it is the union of connected components of $f$-fibres. A saturated neighborhood of an arc of definite points is a product of a disc with the arc, the $f$-fibre components of which are the concentric circles as well as the center of the disc at each point of the arc. A saturated neighborhood of an arc of simple indefinite points is a product of the arc with either a disc with two holes or a Möbius band with one hole (see §1.3).

For saturated neighborhoods of the non-simple points, see §1.4.

Essentially as a device for describing how the connected components of the $f$-fibres cohere to form the manifold, $M$, we introduce the auxiliary space $W_f$, defined by identifying points of $M$ which belong to the same component of the $f$-fibre. We let $q : M \longrightarrow W_f$ be the identification map and $\bar{f} : W_f \longrightarrow \mathbb{R}^2$ be the map defined by $f = \bar{f} \circ q$. This factorization of $f$ into the composition with a map with connected fibres, $q$, and one with finite fibres, $\bar{f}$, is known in algebraic geometry as the Stein factorization [Hart, p. 280]. This identification space was also used by Burlet and de Rham [B-de R] in the special case of stable maps of 3-manifolds into the plane having only definite fold singularities. In that case, $W_f$ can be given the structure of a smooth surface with boundary, the boundary being the q-image of the definite fold curves.

For a general stable f, the local descriptions of $W_f$ in the neighborhood of the q-image of points of S(f) are detailed in §1.4. We give three examples: the drawings show a q-fibre and a neighborhood in $W_f$ of the q-image of that q-fibre

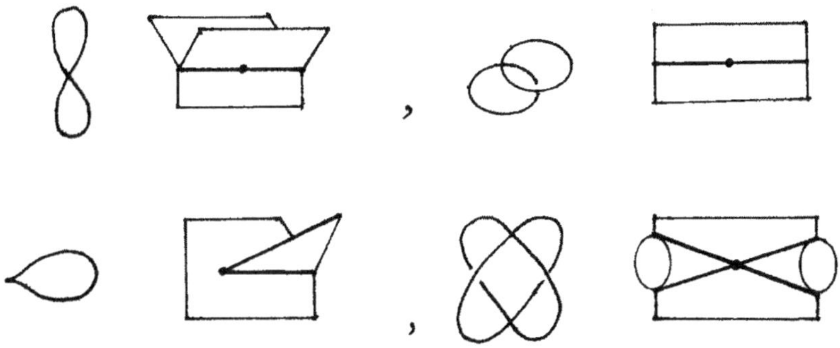

In all of the drawings of neighborhoods in $W_f$, the dark lines are the q-images of the arcs of S(f) in the neighborhood and in each case the heavy point is the q-image of the pictured q-fibre.

Chapter I is devoted to the description of stable maps, their fibres, the auxiliary space $W_f$ and a decomposition of M following the pattern given by the obvious stratification of $W_f$.

In Chapters II and III we restrict ourselves to the M-orientable case. In Chapter II we construct coordinate systems that yield simple canonical forms for the coordinate expressions of f in tubular neighborhoods of arcs of S(f) and in neighborhoods of cusps and double points of $f \mid S(f)$. The proof of the principal technical result of this chapter from which everything else follows is rather long and is therefore deferred to an appendix. The final chapter makes extensive use of the results of this chapter.

In the final chapter we address the problem: *Given a stable map* $f : M \longrightarrow \mathbb{R}^2$ *for* M *a compact orientable* 3-*manifold, classify the* f-*regular homotopy classes of immersions* $(f,h) : M \longrightarrow \mathbb{R}^2 \times \mathbb{R}^2$, *where* (f,h) *and* (f,h') *are* f-*regular homotopic if there is a homotopy,* H, *joining* h *to* h' *such that* $(f,H_t)$ *is an immersion for each* t.

In case N is a compact surface, the necessary and sufficient condition for lifting a stable map $g : N \longrightarrow \mathbb{R}^2$ to an immersion in

$\mathbb{R}^3$ is that each component of $S(g)$ with an even (odd) number of cusps has an orientable (non-orientable) neighborhood. This result of Haefliger [Haef] has been generalized by Blank and Curley [B-C] to lift stable maps (admitting only folds and cusps) between manifolds of the same dimension to an immersion into a line bundle over the target. The problem of classifying all immersions that lift a given map has, as far as I know, been studied only in the case that both the source and target manifolds are 2-dimensional [B, F-T].

The general idea of our classification involves the following: If $(f,h)$ is an immersion of $M$ into $\mathbb{R}^2 \times \mathbb{R}^2$, $h$ must immerse the fibres of $f \mid M - S(f)$ in $\mathbb{R}^2$ and the total turning of the $Th$-image of the tangent to such fibre arcs or circles can be computed. Furthermore $Th$ at $S(f)$ has rank two in the kernel of $Tf$. Thus the total turning of the $Th$-images of those kernel planes along the arcs of $S(f)$ can also be computed. Although neither of these functions is invariant under $f$-regular homotopy, there is a geometrically-defined equivalence relation on the pairs of such functions whose equivalence classes are invariants of $f$-regular homotopy. Furthermore the kernel bundle of $Tf$ along $S(f)$ can be oriented and another $f$-regular homotopy invariant is the orientation preserving or reversing of $Th$ on this bundle.

In addition to the above mentioned invariants which give information about the germ of $(f,h)$ on $q^{-1}(q(S(f))) = \hat{\Sigma}$, the $f$-regular homotopy class of $(f,h) \mid M - \hat{\Sigma}$ is determined by the homotopy class of the map which assigns to each point $P \in M - \hat{\Sigma}$, the unit vector in $\mathbb{R}^2$ in the direction of the $Th$-image of any vector at $P$ orienting the $q$-fibre.

We then give conditions that guarantee that the information so far recorded about the germ of $(f,h)$ at $\hat{\Sigma}$ and on $(M-\hat{\Sigma})$, can arise from an immersion $(f,h)$ on all of $M$. Finally we complete the classification with an invariant that distinguishes among those $(f,h)$ which agree on a neighborhood of $\hat{\Sigma}$ and whose restrictions to the complement of the neighborhood of $\hat{\Sigma}$ are $f$-regularly homotopic.

I have begun Chapters II and III with introductions which will, I hope give accessible resumés of their contents. Some long and technical proofs have been deferred to the Appendix. A Reader's Guide to Notation and a Subject Index are included after the Appendix.

Although this work is the continuation of that begun with León Kushner and Paulo Porto [K, K-L-P], I have tried to make it self contained. I wish to acknowledge here with gratitude the precious contributions to the early development of this work that were made by León Kushner and Paulo Porto.

Note: If $X$ and $Y$ are sets, $(X,Y)$ will denote the set of functions from $X$ to $Y$. In general, functions between manifolds will be assumed to be smooth, i.e. $C^\infty$.

## CHAPTER I

### The Singularities of Stable Maps of Three Manifolds into the Plane



### 1.1 Stable Maps

In this paragraph we describe the stable maps from three dimensional manifolds into the plane. Throughout, we will denote by $M$, a compact orientable 3-manifold without boundary. Let $C(M,\mathbb{R}^2$ be the smooth maps of $M$ into $\mathbb{R}^2$. For $f \in C(M,\mathbb{R}^2)$, the singular set of $f$, $S(f) = \{x \in M \mid \text{rank } Tf(x) < 2\}$. The stable maps $\mathcal{S}(M,\mathbb{R}^2) \subseteq C(M,\mathbb{R}^2)$ are those whose multijet extensions satisfy the usual transversality conditions [Mather, $G^2$]. We give an equivalent description of $\mathcal{S}(M,\mathbb{R}^2)$.

<u>Definition</u>. $f \in \mathcal{S}(M,\mathbb{R}^2)$, if near each point $P \in S(f)$, and in some local coordinates centered at $P$ and $f(P)$, $f$ is one of the following:

$L_0$) $(u,x,y) \longrightarrow (u,x^2+y^2)$, definite fold point or fold point.

$L_1$) $(u,x,y) \longrightarrow (u,x^2-y^2)$, indefinite fold point or saddle point.

$L_2$) $(u,x,y) \longrightarrow (u,y^2+ux-x^3/3)$, cusp point.

In addition the following global conditions are satisfied:

$G_1$) If $P$ is a cusp point, then $\{P\} = f^{-1}(f(P)) \cap S(f)$.

$G_2$) $f \mid S(f) - \{\text{cusps}\}$ is an immersion with normal crossings. (In particular $f \mid S(f)$ has no triple points.)

We let $S_0 = \{\text{definite fold points}\}$, $S_1 = \{\text{indef. fold points}\}$, $C = \{\text{cusps}\}$.

As an immediate consequence of these conditions, we see that for $f \in \mathcal{S}(M,\mathbb{R}^2)$, $S(f)$ is a finite disjoint union of embedded circles in M and C is a finite set of points. On any component of $S(f)$, the points of C separate arcs belonging to $S_0$ from those of $S_1$. Hence on every component of $S(f)$ there is an even number of cusps.

A component of $S(f) - C$ is in either $S_0$ or $S_1$ and is called <u>definite</u> or <u>indefinite</u>, accordingly.

From now on we will write $S$ for $S(M,\mathbb{R}^2)$.

## 1.2 Coordinatized Product Neighborhoods (CPN)

Let $f \in S$. Here we show that each point, p, of M is contained in a neighborhood which is diffeomorphic to the product of an open interval, I, and a surface with boundary, H. The f-image of this neighborhood is diffeomorphic to a rectangle, $I \times J$ where J is a closed interval. If $\phi : I \times H \longrightarrow M$ and $\psi : I \times J \longrightarrow \mathbb{R}^2$ are the diffeomorphisms, then $\psi^{-1} \circ f \circ \phi = 1 \times g : I \times H \longrightarrow I \times J$. We take I and J centered at 0 and assume $\psi(0,0) = f(p)$. For all $u \in I$, $g(u,\cdot)(\partial H) \subseteq \partial J$ and $g(u,\cdot)$ is a Morse function except for $g(0,\cdot)$ in case p is a cusp.

<u>Definition</u>. We say that smooth mappings $f : M \longrightarrow P \longleftarrow Q : g$ <u>meet transversally</u> or f <u>is transverse to</u> g, $f \pitchfork g$ at $z \in P$ if either $z \notin f(M) \cap g(Q)$ or $z \in f(M) \cap g(Q)$, and $Tf(T_xM) + Tg(T_yQ) = T_zP$ for all $x \in f^{-1}(z)$ and $y \in g^{-1}(z)$. We say f and g <u>meet transversally</u> or f <u>is transverse to</u> g if the condition is satisfied at each $z \in P$. If f, g or both are inclusions we say $M \pitchfork g$, $f \pitchfork Q$ or $M \pitchfork Q$ respectively.

<u>Proposition 1</u>. <u>Let</u> $f \in S$, $I = (-1,1)$, $J = [-1,1]$. <u>For each</u> $y \in \mathbb{R}^2$ <u>there is a diffeomorphism</u> $\psi : I \times J \longrightarrow \mathbb{R}^2$ <u>such that</u> $y = \psi(0,0)$ <u>and the composition</u>,

$$f^{-1}(\psi(I\times J)) \xrightarrow{f} \psi(I\times J) \xrightarrow{\psi^{-1}} I \times J \xrightarrow{proj} I$$
$$\underbrace{\qquad\qquad\qquad\qquad\qquad\qquad\qquad\qquad}_{h}$$

<u>is a trivial bundle</u>.

<u>Proof</u>. This is an easy application of Ehresmann's Theorem [E,T] which states:

<u>Let</u> X <u>and</u> Y <u>be connected manifolds without boundary</u>. <u>If</u> f <u>is a proper submersion onto</u> Y, <u>then</u> $f : X \longrightarrow Y$ <u>is a locally trivial fibration</u>. <u>If</u> $\partial X \neq \emptyset$ <u>and</u> f <u>and</u> $f \mid \partial X$ <u>are submersions onto</u>

$Y$ *then* $f : X \longrightarrow Y$ *is a locally trivial fibration with* $\partial(f^{-1}(y)) = (f|\partial X)^{-1}(y)$ *for all* $y \in Y$.

If $y \in \text{regval } f = \{\text{regular values of } f\}$, we may take $\psi$ as any diffeomorphism whose image lies completely in regval $f$ and the proposition is immediate.

If $y \in f(S(f))$, let $\psi : (0\times J) \longrightarrow \mathbb{R}^2 : (0,0) \longrightarrow y$ be any diffeomorphism transverse to $f$ at $y$ such that $\psi(0\times J) \cap f(S(f)) = \{y\}$. Extend $\psi$ to a diffeomorphism of $I \times J$ so that for each $u \in I$, $\psi \mid (u\times J) \pitchfork f$, and such that $\partial\psi(I\times J) = \psi(I\times\partial J) \subseteq \text{regval } f$. Thus both $\psi$ and $\psi \mid I \times \partial J$ are transverse to $f$, and so both $f^{-1}(\psi(I\times J))$ and $f^{-1}\psi(I\times\partial J)$ are smooth manifolds. Since $\psi(I\times\partial J) \subseteq$ regval $f$, $f : f^{-1}\psi(I\times\partial J) \longrightarrow \psi(I\times\partial J) \cap f(M)$ is a surjective submersion. Since $\psi(I\times\partial J) \subseteq \text{regval } f$, each component of $\psi(I\times\{-1,+1\})$ must be wholly inside of or wholly outside of $f(M)$. In case $y$ is the image of a cusp or an indefinite point, both components are in $f(M)$ since $\psi(0\times\{-1,+1\}) \subseteq f(M)$. If $y$ is the image of a definite point, only one of $\psi(0\times\{\pm 1\})$ is in $f(M)$ so only one component of $\psi(I\times\partial J)$ is in $f(M)$. However in all cases since at least one component of $\psi(I\times\partial J)$ is in $f(M)$, the composition $\text{proj}\circ\psi^{-1}$ is a surjective submersion onto $I$. Finally, since $f \pitchfork \psi \mid u \times J$ for all $u \in I$, if $y = \psi(u,v) = f(x)$,

$$Tf(T_xM) + T\psi(T_{(u,v)}(u\times J)) = T_y(\mathbb{R}^2)$$

or

$$T(\psi^{-1}\circ f)(T_xM) + T_{u,v}(u\ J) = T\psi^{-1}(T_y\mathbb{R}^2)$$

or

$$Th(T_xM) = T(\text{proj}\circ\psi^{-1})(T_y\mathbb{R}^2) = T_u(I).$$

Thus at each $x \in f^{-1}\psi(I\times J)$, $h$ has rank one. The surjectivity of $h$ is trivial since $h \mid \partial(f^{-1}\psi(I\times J))$ is surjective. //

For any $y \in \mathbb{R}^2$ let $\psi : I \times J \longrightarrow \mathbb{R}^2$ be any diffeomorphism centered at $y$, as in the previous proposition. Since the bundle of the proposition is trivial there is a surface with boundary, $\tilde{T}$ and a diffeomorphism $\phi : I \times \tilde{T} \longrightarrow f^{-1}\psi(I\times J)$ such that $h\circ\phi : I \times \tilde{T} \longrightarrow I$ is just the projection. The commutativity of the following diagram defines $g : I \times \tilde{T} \longrightarrow J$:

$$\begin{array}{ccc} I \times \tilde{T} & \xrightarrow{\ \phi\ } & M \\ {\scriptstyle 1 \times g}\downarrow & & \downarrow{\scriptstyle f} \\ I \times J & \xrightarrow{\ \psi\ } & \mathbb{R}^2 \end{array}$$

For any $x \in f^{-1}(y)$, let $T$ be the connected component of $\tilde{T}$ with $\phi(0,T) \ni x$. Thus we have shown that at each $x \in M$ there is a coordinatized product neighborhood with the following definition.

*Definition*. A *coordinatized product neighborhood* (CPN) *of* $f$ *at* $x$ is a triple of maps $(\phi,\psi,g)$ where $\phi : I \times T \longrightarrow M$ and $\psi : I \times J \longrightarrow \mathbb{R}^2$ are diffeomorphisms into, $g : I \times T \longrightarrow J$ such that $f \circ \phi = \psi \circ (1 \times g)$, $\psi(I \times J)$ is a neighborhood of $f(x)$, and $\phi(I \times T)$ is the connected component of $f^{-1}\psi(I \times J)$ containing $x$. If $f(x) \in \psi(u_0 \times J)$, then $\psi(u_0 \times J) \cap f(S(f)) = \{f(x)\} \cap f(S(f))$. Here $I$ $(J)$ are open (closed) intervals and $T$ is a surface with boundary. We will sometimes refer to $I \times T$ or $\phi : I \times T \longrightarrow M$ as a product neighborhood of $x$.

Obviously, for $y \in \mathbb{R}^2$, the $\psi$ given in the proposition is not unique, the only important property of $\psi$ is that $\phi \mid (u \times J) \pitchfork f$ for all $u \in J$ and that $\psi((I \times \partial J) \cup (0 \times J - 0)) \subseteq \operatorname{regval} f$. Notice that if $x'$ is in the connected component of $f^{-1}(f(x))$, then a CPN of $f$ at $x$ is also one at $x'$.

*Definition*. For any $y \in \mathbb{R}^2$, an embedding of a (closed) interval $\alpha : J \longrightarrow \mathbb{R}^2$ is called *a* *(closed)* *transverse* *arc* at $y$ if $y$ is interior to $\alpha(J)$, $\alpha \pitchfork f$, and $\alpha(J) \cap f(S(f)) = \{y\} \cap f(S(f))$. For $x \in M$, if $\alpha : J \longrightarrow \mathbb{R}^2$ is a transverse arc at $f(x)$, then the component of $f^{-1}(\alpha(J))$ containing $x$ is called a *transverse manifold at* $x$.

Given any $y \in \mathbb{R}^2$, there are many transverse arcs at $y$. However, if $\alpha_0$ and $\alpha_1$ are transverse arcs at $y$ both of which begin and end in the same components of $\mathbb{R}^2 - f(S(f))$, then there is a smooth family $\alpha_t : J \longrightarrow \mathbb{R}^2$, $t \in [0,1]$ of transverse arcs at $y$, joining $\alpha_0$ to $\alpha_1$. Call two transverse arcs *equivalent* if their end points are in the same components of $\mathbb{R}^2 - f(S(f))$. It is easy to see that:

*Proposition 2*. *Any two transverse manifolds at* $x_0 \in M$ *over equivalent transverse arcs at* $f(x_0)$ *are diffeomorphic*.

Proof. Let $\alpha_0, \alpha_1 : J \longrightarrow \mathbb{R}^2$ be equivalent transverse arcs at $y_0 = f(x_0)$. Let $I_\varepsilon = (-(1+\varepsilon), 1+\varepsilon)$ and let $A : I_\varepsilon \times J \longrightarrow \mathbb{R}^2$ be a family of transverse arcs at $y_0$ such that $A(0,\cdot) = \alpha_0(\cdot)$, $A(1,\cdot) = \alpha_1(\cdot)$. Denote $A(t,\cdot)$ by $\alpha_t(\cdot)$, $t \in I_\varepsilon$. For all $t \in I_\varepsilon$, $\alpha_t(0) = y_0$. The map $1 \times A : I_\varepsilon \times J \longrightarrow I_\varepsilon \times \mathbb{R}^2 : (t,s) \longrightarrow (t, \alpha_t(s))$ is an embedding. Let $N = (1\times A)(I_\varepsilon \times J)$. Let $F = 1 \times f : I_\varepsilon \times M \longrightarrow I_\varepsilon \times \mathbb{R}^2 : (t,x) \longrightarrow (t, f(x))$. Since $f \pitchfork \alpha_t$ for all $t \in I_\varepsilon$, $F \pitchfork N$ and $F \pitchfork \partial N$. Thus $F^{-1}(N) = P$ is a smooth manifold with smooth boundary $F^{-1}(\partial N)$. It is easy to check that the maps

$$G = \text{proj} \circ F : P \xrightarrow{F} I_\varepsilon \times \mathbb{R}^2 \xrightarrow{\text{proj}} I_\varepsilon,$$

and $G \mid \partial P$ are proper and are surjective submersions. Hence all the fibres of $G$ are diffeomorphic. But for $t \in I_\varepsilon$, $G^{-1}(t) = F^{-1}(t, \alpha_t(J)) = (t, f^{-1}(\alpha_t(J)))$. Thus all of the transverse manifolds over the family are diffeomorphic. //

Thus the diffeomorphism classes of transverse manifolds at $x \in M$ correspond to equivalence classes of transverse arcs at $y = f(x) \in \mathbb{R}^2$. Every transverse manifold at $x \in M - S(f)$ is diffeomorphic to $J \times S^1$ and we can choose a CPN at $x$ so that $1 \times g : I \times J \times S^1 \longrightarrow I \times J$ is just the projection. If $x \in S(f)$ is not a double point of $f \mid S(f)$, all transverse manifolds at $x$ are diffeomorphic. If $x$ is a double point of $f \mid S(f)$, there are two equivalence classes of transverse arcs at $f(x)$, since the arcs of $f(S(f))$ cut $\mathbb{R}^2$ into four pieces near $f(x)$. In the next two sections we determine all transverse manifolds at points of $S(f)$.

## 1.3 The space $W_f$ and transverse manifolds at simple points

Let $f \in S$. Define $W_f$ as the quotient of $M$ obtained by identifying two points of $M$ if they are in the same connected component of the same fibre of $f$. Let $q_f : M \longrightarrow W_f$ be the quotient map and let $\bar{f} : W_f \longrightarrow \mathbb{R}^2$ be defined by $f = \bar{f} \circ q_f$. Henceforth when no confusion is likely we will omit the subscript, $f$.

Proposition 1. *If* $x \in S(f)$, *then* $q^{-1}(q(x)) \cap S(f) = \{x\}$ *if* $x \in S_0 \cup C$ *and if* $x \in S_1$, $q^{-1}(q(x)) \cap S(f) = \{x\}$ *or* $\{x, x'\}$.

Proof. Condition $G_1$) guarantees that $q^{-1}q(x) \cap S(f) = \{x\}$ if $x \in C$ and condition $G_2$) guarantees that $f \mid S(f)$ has at most double points.

From $L_0$), the local form of $f$ about a point of $S_0$, we see that in local coordinates the f-preimage of a point $(U,Y) \in \mathbb{R}^2$ is just $\{(U,x,y) \mid x^2 + y^2 = Y\}$ so the preimage of the origin is just the origin, which is a connected component of the fibre. //

Definition. A point $x \in S(f)$ for which $q^{-1}q(x) \cap S(f) = \{x\}$ is called a simple singular point. By the preceding proposition if $x \in S(f)$ is not simple, then $x$ must be in $S_1$.

Lemma 1. *If* $f \in S$, *then* $f \mid M - S_0$ *is an open map.*

Proof. On $M - S(f)$, $f$ is a submersion, hence open. At a point of $S_1$, $f$ is open since $(x,y) \longrightarrow (x^2-y^2)$ is an open map. At a cusp point in local coordinates (see $L_2$)) the image of $(u,x,y)$ in $(-\delta,\delta) \times (-\varepsilon,\varepsilon) \times (-\eta,\eta)$ is just the region between the graphs of $u \longrightarrow (u\varepsilon-\varepsilon^3/3)$ and $u \longrightarrow (-u\varepsilon+\eta^2+\varepsilon^3/3)$ for small enough $\delta < \varepsilon^2/3$. //

Remark. As already noted, a (CPN) of $f$ at $x$ is also a (CPN) of $f$ at any point $x' \in q^{-1}q(x))$. Thus it makes sense to speak of a (CPN) of $f$ at $w \in W$. We will use either designation.

Lemma 2. *Given a CPN,* $(\phi,\psi,g)$ *for* $f$ *at* $x$, *we can define a homeomorphism into,* $\vartheta : W_{1\times g} \longrightarrow W_f$ *so that the diagram commutes:*

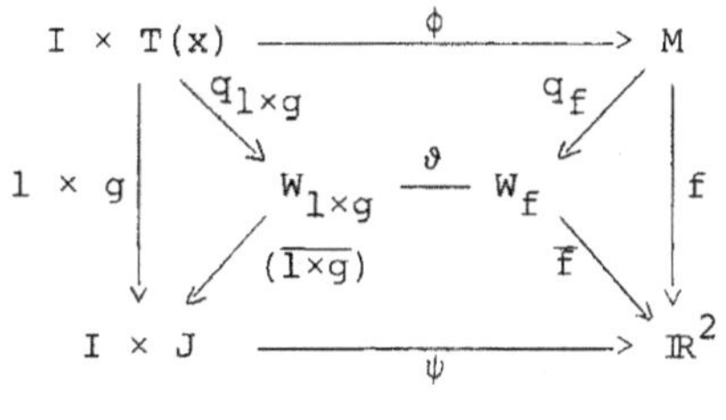


Proof. Let us write $q'$ for $q_{1\times g}$ and $W'$ for $W_{1\times g}$. For any $(u,z)$ $I \times T(x)$, let $\vartheta(q'(u,z)) = q_f(\phi(u,z))$. To see that this defines a map $\vartheta$, let $q'(u,z') = q'(u,z)$, then we know $g(u,z') = g(u,z)$, and $(u,z)$ and $(u,z')$ are in the same connected component, $K$ of $(1\times g)^{-1}(u,g(u,z))$. Thus we know that $f(\phi(u,z)) = f(\phi(u,z')$ and $\phi(u,z)$ and $\phi(u,z')$ are in $\phi(K)$ a connected component of $f^{-1}(f(\phi(u,z))$. So $q_f(\phi(u,z)) = q_f(\phi(u,z'))$. So $\vartheta$ is well defined and makes the diagram commute. Since both $W'$ and $W_f$ have the quotient topology, $\vartheta$ is obviously a homeomorphism into. //

We write $\Sigma_f$ for $q(S(f))$ and $\hat{\Sigma}_f$ for $q^{-1}(\Sigma_f)$. As usual we will generally omit the subscript $f$.

*Proposition 2.* $W - \Sigma$ *is a smooth (non-compact) surface, immersed by* $\bar{f}$ *in* $\mathbb{R}^2$, *and* $q : M - \hat{\Sigma} \longrightarrow W - \Sigma$ *is a circle bundle, trivial if* $M$ *is oriented.*

*Proof.* Since $\bar{f} \mid W - \Sigma$ is a local homeomorphism, the first part of the proposition results from giving $W - \Sigma$ the $\bar{f}$-pulled back smooth structure. For the second part, we noted at the end of the previous paragraph that for any $x \in M - \hat{\Sigma}$, any transverse manifold is diffeomorphic to $J \times S^1$, and a CPN of $f$ at $x$, $(\phi, \psi, g)$ can be chosen so that $1 \times g : I \times J \times S^1 \longrightarrow I \times J$ is just the projection. Since $W_{1\times g} = I \times J$, the local trivialization of $q \mid M - \hat{\Sigma}$ above $\vartheta(I\times J)$ is given by Lemma 2:

$$\begin{array}{ccc} I \times J \times S' & \xrightarrow{\quad\phi\quad} & M - \hat{\Sigma} \\ {\scriptstyle \mathrm{proj} = q_{1\times g}}\downarrow & & \downarrow{\scriptstyle q} \\ I \times J & \xrightarrow{\quad\vartheta\quad} & W - \Sigma . \end{array}$$

The bundle is trivial if $M$ and hence $M - \hat{\Sigma}$ are orientable. //

*Proposition 3.* *Let* $x \in S_0$. *The transverse manifold,* $T(x)$, *of* $f$ *at* $x$ *is diffeomorphic to a disc. We can choose a CPN of* $f$ *at* $x$ *of the form*

$$\begin{array}{ccc} I \times \mathbb{D} & \xrightarrow{\quad\phi\quad} & M \\ {\scriptstyle 1 \times g}\downarrow & & \downarrow{\scriptstyle f} \\ I \times J & \xrightarrow{\quad\psi\quad} & \mathbb{R}^2 \end{array}$$

where $g(t,x,y) = x^2 + y^2$; where $\mathbb{D} \subseteq \mathbb{R}^2$ is a disc about $0$.

*Proof.* This is just a restatement of condition $L_0$) on $f \in S$. //

*Proposition 4.* *Let* $x$ *be a simple point of* $S_1$. *There is a CPN of* $f$ *at* $x$, $(\phi, \psi, g)$ *such that* $1 \times g : I \times T(x) \longrightarrow I \times J$ *is a product map, where* $g : T(x) \longrightarrow J$ *is surjective with a single saddle singularity.* i) *If* $T(x)$ *is orientable,* $T(x)$ *is diffeomorphic to a disc with two holes and* $W_g$ *is a "Y-shaped" graph.* ii) *If* $M$ *is not orientable and* $T(x)$ *is not orientable, it is diffeomorphic to a Möbius band with one hole and* $W_g$ *is just an interval.*

*Proof.* Given any CPN of $f$ at $x$, $(\phi', \psi', g')$. By virtue of the local coordinate form of $f$, $L_1$), we know that for each $u \in I$, $g'_u : T(x) \longrightarrow J$ is stable with one saddle singularity whose image is $0$ in $J$. Further since $g'_u$ maps the boundary of $T(x)$ to the boundary of $J$, $g'_u$ is constant in $u$ on $\partial T(x)$. Since $g'_u$ are stable for all $u$, it is easy to construct a diffeomorphism, $1 \times \gamma :$ $I \times T(x) \circlearrowleft$ such that $(\phi = \phi' \circ (1\times\gamma),\ \psi = \psi',\ 1 \times g = (1\times g') \circ (1\times\gamma))$ is a CPN of $f$ at $x$ and $g : T(x) \longrightarrow J : z \longrightarrow g(z)$. (See [$G^2$], Theorem 3.3, p. 12.4 or [$L_1$], Theorem 12.3, p. 73.)

If $T(x)$ is orientable it is diffeomorphic to a disc with two holes (see, for example, [H], §3, p. 201). The $g : T(x) \longrightarrow J$ factors through $W_g$ as $\bar{g}\circ q_g$ which we picture below:

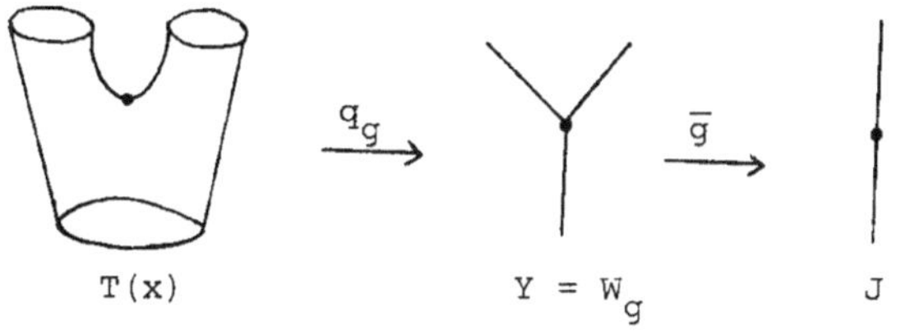


Since $1 \times g : I \times T(x) \longrightarrow I \times J$ is a product map, $W_{1\times g}$ is also a product of $I \times Y$, where $Y$ is so named for obvious reasons.

If $M$ were not assumed orientable, there would be another possible $T(x)$, namely a Möbius band with a hole in it and in that case $W_g = J$. The factorization in the case of $g \mid T(x) = \bar{g}\circ q_g$ is shown below.

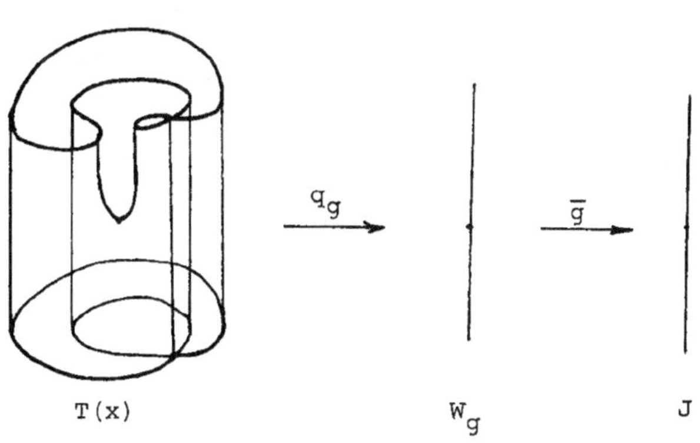

We picture $T(x)$ with $g$-level curves through the saddle in both cases:

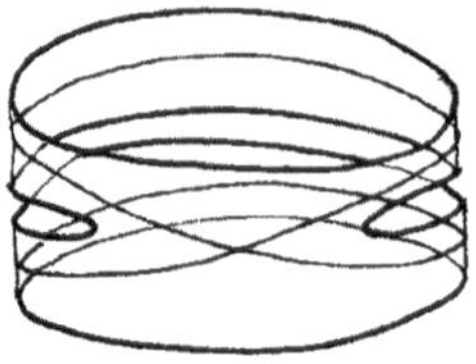

T(x) orientable

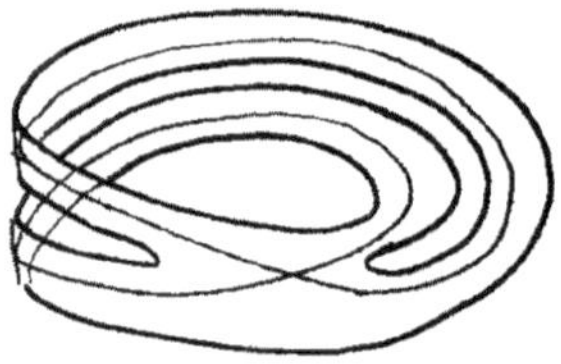

T(x) non-orientable //

*Proposition 5. Let* $x \in C$. *The transverse manifold,* $T(x)$, *of* $f$ *at* $x$ *is diffeomorphic to a disc with one hole. We can choose* $(\phi,\psi,g)$, *a CPN of* $f$ *at* $x$ *such that* $g_u : T(x) \longrightarrow J$ *has no singularities for* $u < 0$, *has a saddle and a minimum for* $u > 0$. *As* $u$ *decreases to* $0$, *the saddle and minimum coalesce. The* $q$*-image of* $u \times T(x)$ *is an interval for* $u \leqslant 0$ *and is a Y-shaped graph for* $u > 0$. *As* $u$ *decreases to zero one of the arms shrinks to a point.*

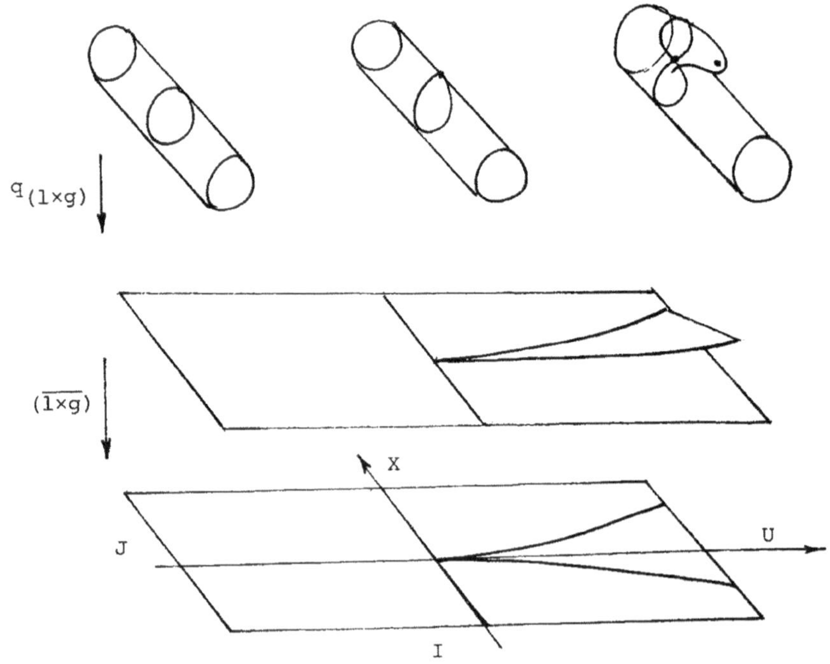

Proof. Here again, all that is needed is to examine the local form $L_2$ at a cusp. In coordinates our map near the cusp is $(u,x,y) \longrightarrow (u, y^2+ux-x^3/3)$. If we label the target coordinates $(U,X)$ and take $I \times J$ as the coordinate rectangle $|U| < \varepsilon$, $|X| < \delta$ with $9\delta^2 > 4\varepsilon^3$, then the image of the singular set is $\{(U,X) \mid 4U^3 = 9X^2\}$.

We take the X-axis as our transverse arc. The singular arcs above $4U^3 = 9X^2$ are $S_1$ points for $X > 0$ and $S_0$-points for $X < 0$. We draw $g_u \mid u \times T(x)$ for $u < 0$, $u = 0$ and $u > 0$.

## 1.4 The non-simple $S_1$-points and their transverse manifolds

Let $T$ be a transverse manifold at $x$, a simple indefinite point. Let $g : T \longrightarrow J$ be the Morse function given in Proposition 4 of §1.3. We know that $g$ has simple saddle singularity on the fibre above one point, say $0 \in J$. The fibres $g^{-1}(s) = T_s$ are circles (one or two) for all $s \neq 0$. The transition from $T_s$ for $s < 0$ to $T_{s'}$ for $s' > 0$ is effected by one surgery--an oriented one in case $T$ is oriented (i), and a non-oriented one in case $T$ is not oriented (ii):

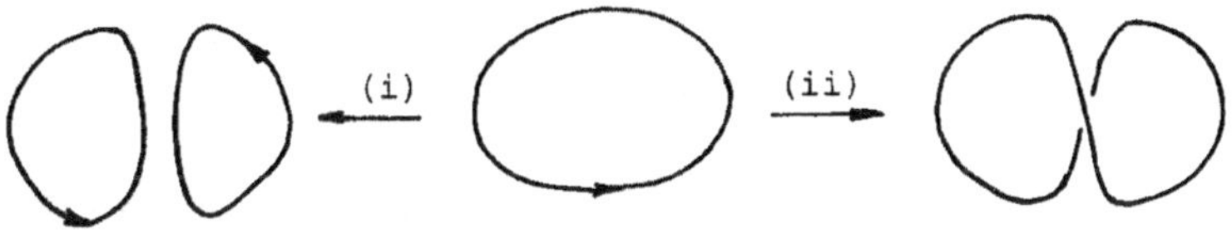


To describe the transverse manifolds through non-simple points we must consider transitions through fibres resulting from two simultaneous surgeries.

Let $x$ be a non-simple indefinite point; we know that $g^{-1}g(x) \cap S(f) = \{x,x'\}$. Since $f(x) = f(x')$ and $f \mid S(f)$ has only normal crossings by $G_2$) of 1.1, we may construct a CPN of $f$ at $x$, $(\phi,\psi,g)$ such that the image of $S(1\times g)$ in $I \times J$ cuts the rectangle into four regions as illustrated.

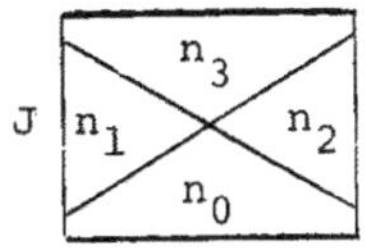

In each region, we let $n_i$ be the number of components of the fibre of $(1\times g) : I \times T \longrightarrow I \times J$ above any point of that region. Furthermore it is no restriction to assume that we've chosen our CPN so that $n_0 \leqslant n_i$ for all $i$. This condition is satisfied by choosing one of the classes of transverse arcs mentioned at the end of §1.2 in the construction of our CPN.

Proposition 1. The non-simple indefinite points are classified by the following five arrays of numbers.

(i) In the orientable case 

(ii) In the non-orientable case 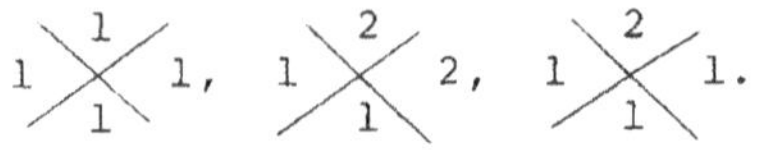


Proof. We first check that $n_0 = 1$. We follow the fibre above arcs which cross from the $n_0$-region to the $n_1$-region and to the $n_2$-region. The change in the fibres as we travel across the image of $S(1\times g)$ is produced by a single surgery for each of these paths. Each of these surgeries involves a single circle above the $n_0$-region. For if not, if the surgery required to make the transition across the $n_0 - n_1$ border (for instance) involved two circles from the $n_0$-region, then $n_1 = n_0 - 1$ contradicting the minimality of $n_0$. Thus $n_0 = 1$ or $= 2$. The transition from the $n_0$-region to the $n_3$-region results from performing both surgeries. If $n_0 = 2$, the $(1\times g)$ preimage of the transverse arc through the crossing (i.e. the transverse manifold through $\phi^{-1}(x)$) would not be connected--contradiction. Thus $n_0 = 1$.

We know the transition at a single indefinite point results from an oriented or an unoriented surgery where an oriented surgery replaces $c$  $\bar{c}$ by  and an unoriented surgery replaces $c$ 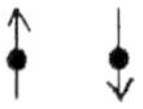 $c^*$ by . Orient the circle fibre above the $n_0$-region arbitrarily. Label the $1 \times g(S(1\times g))$ arcs in $I \times J$ as $A$ and $B$ and label the pairs of points on the circle fibre at which the surgeries occur when crossing arcs $A$ (or $B$) with the letters $A$, $\bar{A}$ (or $B$, $\bar{B}$) for an oriented surgery and $A$, $A^*$ (or $B$, $B^*$) for a non-oriented surgery. We display all cases below:

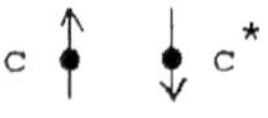

(i) Both surgeries oriented.

$(i_a)$

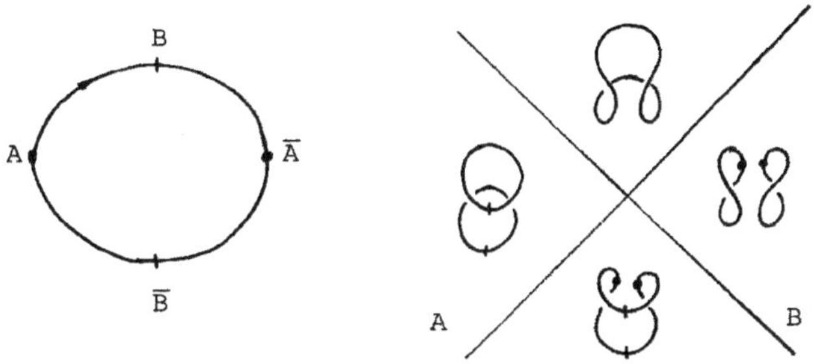


$(i_b)$

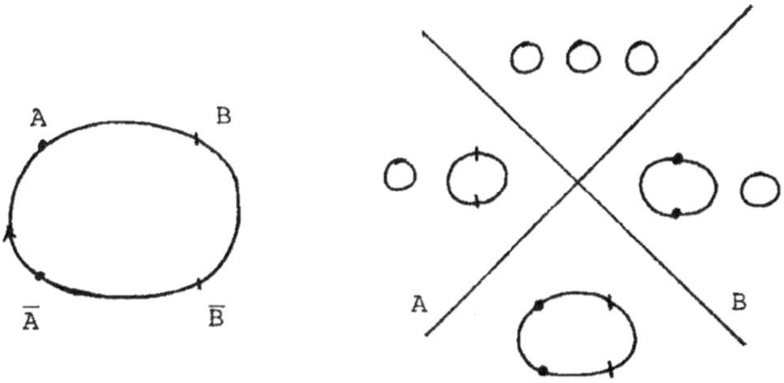


(ii) One or both surgeries unoriented.

$(ii_a)$

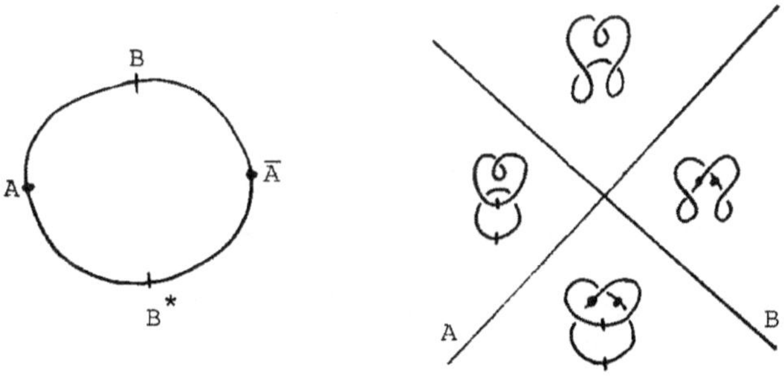


Notice that this is the same as two unoriented surgeries only exchanging the roles of $I$ and $J$ in the target by a clockwise rotation of $\pi/2$.

$(ii_b)$

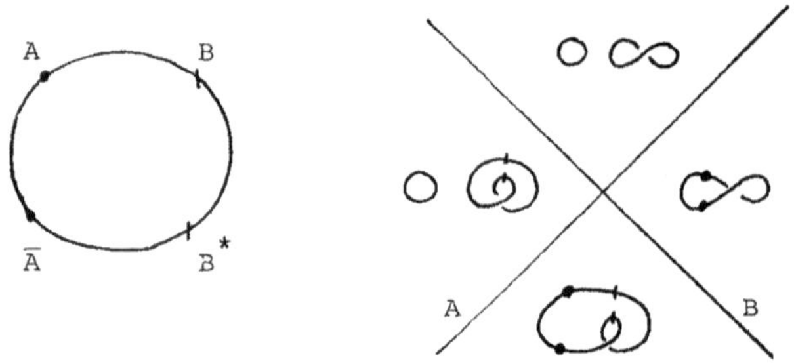


$(ii_c)$

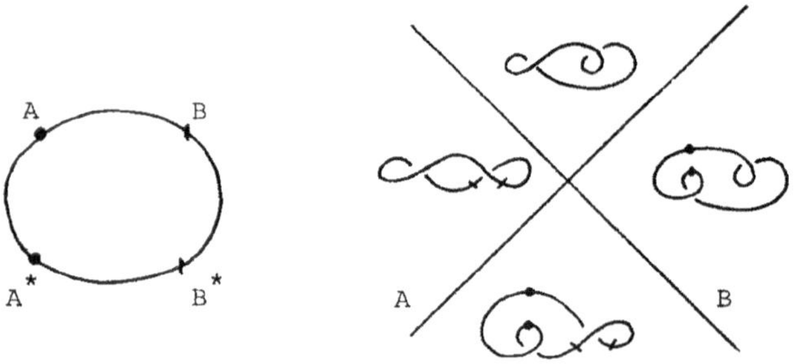


We will refer to non-simple points and their q-images as (1·2·2·1), (1·2·2·3), (1·2·1·1), (1·2·1·2) and (1·1·1·1) points. As noted at the end of §1.2 there are two classes of transverse manifolds at non-simple points. We now choose one. At an (a·b·c·d)-non-simple point, by a transverse manifold we will mean a manifold above an arc through the crossing joining the 'a' and 'd' regions. With this convention, we know that a transverse manifold at an (a·b·c·d) point has a + d boundary circles and has a Morse function constant on the boundary components having precisely two singular points both of which are saddles. Since we also know that the (1·2·2·1) and (1·2·2·3) transverse manifolds are orientable and the other three are not, we know the diffeomorphism type of the transverse manifolds:

*Corollary. The diffeomorphism type of the transverse manifolds at the non-simple indefinite points are:*

- *A torus with two holes at a* (1·2·2·1)-*point.*
- *A disc with three holes at a* (1·2·2·3)-*point.*
- *A Klein bottle with two holes at a* (1·1·1·1) *or* (1·2·1·1)-*point.*
- *A Möbius band with two holes at a* (1·2·1·2)-*point.*

<u>Remarks</u>.

1. The particular pictures of the fibres in $I \times T$ drawn above were chosen because it is easy to visualize them as level sets of a Morse function on $T$. We illustrate this for the transverse manifold over the arc through the crossing. We draw in the fibre over the crossing point.

$i_a)$ $(1\cdot2\cdot2\cdot1)$

$i_b)$ $(1\cdot2\cdot2\cdot3)$

$ii_a)$ $(1\cdot2\cdot1\cdot1)$

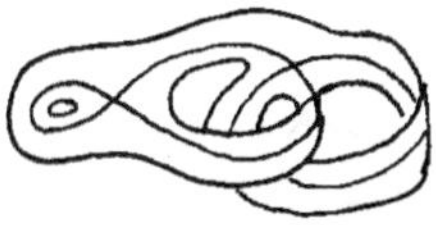

$ii_b)$ $(1\cdot2\cdot1\cdot2)$

$ii_c)$ $(1\cdot1\cdot1\cdot1)$

2. Notice that although at each cusp and non-simple indefinite point we have product neighborhoods, the map $1 \times g : I \times T \longrightarrow I \times J$ is not a product map and $W_{1\times g}$ is <u>not</u> of the form $I \times Z$.

We now know all possible local descriptions of $W_f$ for $f$ a stable map of a compact 3-manifold into a 2-manifold. We give a table of all $q$-images of product neighborhoods.

<u>Definition</u>. For any $Q \in W$, we call the $q$-image of any product neighborhood of any point in $q^{-1}(Q)$ a <u>conical neighborhood of</u> $Q$.

This definition is motivated by the fact that the $q$-image of any product neighborhood of any point in $q^{-1}(Q)$ is homeomorphic to the join of $Q$ with the boundary of the $q$-image.

| If $q(x) = Q \in$ | The q-image of a product neighborhood of x is | T(x) is a |
|---|---|---|
| $W - q(S(f))$ |  | $(\text{disc})_{(1)}$ |
| $q(S_0)$ |  | disc |
| $q(S_1)$, simple |  | $(\text{disc})_{(2)}$ |
| | 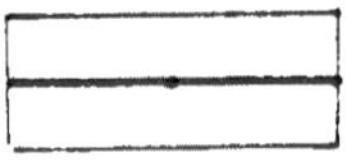 | $(\text{Möbius})_{(1)}$ |
| $q(C)$ | 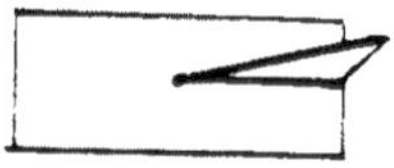 | $(\text{disc})_{(1)}$ |
| $q(S_1)$ non-simple of type | | |
| (1·2·2·1) | 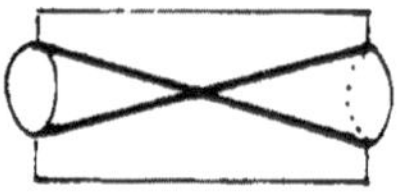 | $(\text{Torus})_{(2)}$ |
| (1·2·1·1) | 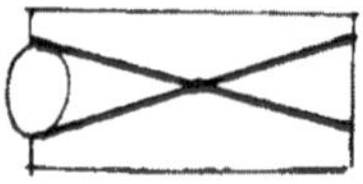 | $(\text{Klein})_{(2)}$ |
| (1·2·2·3) | 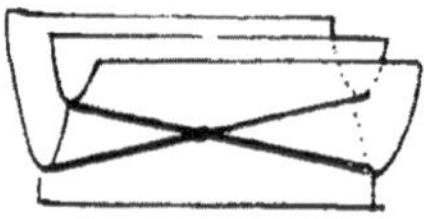 | $(\text{disc})_{(3)}$ |
| (1·1·1·1) | 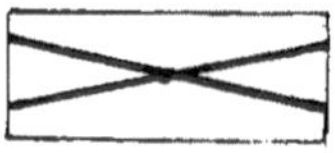 | $(\text{Klein})_{(2)}$ |
| (1·2·1·2) | 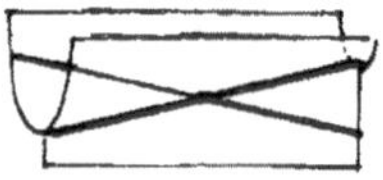 | $(\text{Möbius})_{(2)}$ |

The heavy lines in the figures represent the q-images of S(f) near Q. Here the subscript in parentheses in the description of T(x) denotes the number of discs deleted: e.g. $(\text{Möbius})_{(k)}$ means a Möbius band with k discs removed.

## 1.5 A partial orientation of $S(f)$

We assign to each connected component $\Omega$ of $\mathbb{R}^2 - f(S(f))$ an integer, $n_f(\Omega)$, the number of connected components of the fibre of $f$ above any point of $\Omega$. Obviously $\overline{f} : \overline{f}^{-1}(\Omega)$ is a covering space and $n_f(\Omega)$ is the number of times $\overline{f}^{-1}(\Omega)$ covers $\Omega$. Every arc of $f(S(f)) - f(C \cup \{\text{double points}\})$ is contained in the boundary of two components of $\mathbb{R}^2 - f(S(f))$. If these components have different $n_f$-values, orient the arc so that the region with the larger $n_f$-value is on the left. This partial orientation on $f(S(f))$ induces orientations on some of the arcs of $S(f) - C \cup \{\text{double points}\}$ and on $q(S(f) - C \cup \{\text{double points}\})$. Clearly all arcs of $S_0$ are given an orientation; locally the image of $M$ lies to the left of an oriented immersed arc of $S_0$. Also any arc of simple points of $S_1$ whose transverse manifold, $T$, is a disc with two holes is oriented by the convention; the arms of the $Y$ shaped $q(T)$ are mapped to the left of the image of the arc. However no orientation is given to those arcs of simple points of $S_1$ whose transverse manifolds are Möbius bands with one hole. We examine these orientations at the vertices:

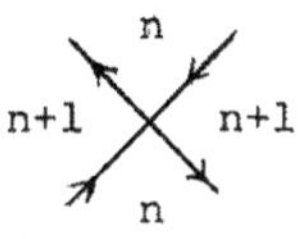


1·2·2·1

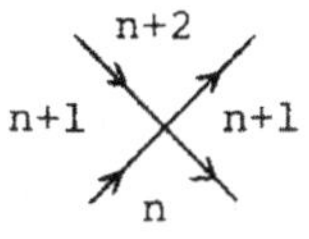


1·2·2·3

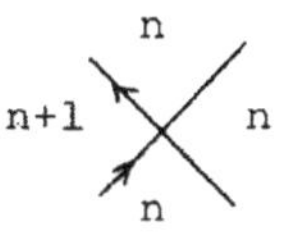


1·2·1·1

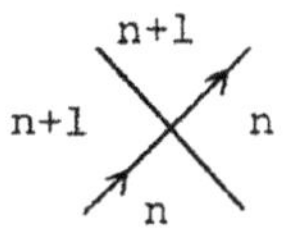


1·2·1·2

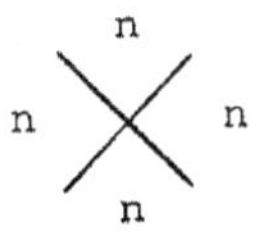


1·1·1·1

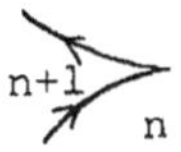


cusp

Examining the situations that arise if $M$ is orientable gives:

*Proposition. If $M$ is orientable:*

1) *Each component of* $S(f) - \{(1{\cdot}2{\cdot}2{\cdot}1)\text{-points}\}$ *can be oriented so that the value of* $n_f$ *is always greater to the left of the image than to the right.*

2) *On every component of* $S(f)$ *there is an even number of* $(1{\cdot}2{\cdot}2{\cdot}1)$ *points.*

We will henceforth refer to this orientation as the standard orientation (on arcs of $S(f)$ and their $q$- and $f$-images as well).

Using the standard orientation we distinguish two cusps that arise:

Definition. We designate a cusp as a (+1)-cusp if the indefinite arc of $S(f)$ that abuts the cusp ends at the cusp and the definite arc that abuts the cusp starts there. If the roles of the definite arcs are reversed we call the cusp a (-1)-cusp.

## 1.6 Decomposing $M$ and $W_f$

In this section we decompose $M$ and $W = W_f$ into simple pieces. We begin with some notation. Let $\Sigma = q(S(f))$, and let the set of components of $W - \Sigma$ be denoted by $\mathcal{R}$. Each $R \in \mathcal{R}$ is a surface with boundary (possibly with corners) and $\bar{f} \mid R$ is an immersion. $B(R) = q^{-1}(R)$ is a circle bundle over $R$, trivial if $M$ is orientable.

In $\Sigma$, let $V$ be the set of $q$-images of cusp points and non-simple points of $S_1$. We will refer to $V$ as the vertices of $W$, a vertex $v$ is called a non-simple vertex (a cusp vertex) if $q^{-1}(v)$ meets $S(f)$ in two points ($q^{-1}(v)$ meets $S(f)$ at a cusp). Let $\mathcal{C}$ be the set of components of $\Sigma - V$. An arc $c \in \mathcal{C}$ is immersed by $\bar{f}$ (with double points at worst). An element $v \in V$ is the end point of at most four distinct arcs of $\mathcal{C}$; a cusp being the end point of precisely two arcs in $\mathcal{C}$. We write $\mathcal{C} = \mathcal{C}_0 \cup \mathcal{C}_1$ where $\mathcal{C}_i$ is the set of $q$-images of components of $S_i$ - {non-simple points}.

About each $v \in V$, choose a conical neighborhood, $N(v)$ (see the end of §1.4). We know that $B(v) = q^{-1}(N(v))$ is diffeomorphic to an interval times the transverse manifold $T(v)$ at $v$.

Let $c \in \mathcal{C}$ and let $N_{\bar{f}}$ be the normal bundle of the immersion $\bar{f}|_c : c \longrightarrow \mathbb{R}^2$ and let $p : N_{\bar{f}} \longrightarrow c$ be the bundle projection. We may assume that we have an immersion $e : N_{\bar{f}|_c} \longrightarrow \mathbb{R}^2$ which is an embedding of the fibres and such that $\bar{f}|_c = e \circ$ (zero section).

Proposition. For each $c \in \mathcal{C}$, there is a neighborhood $B(c)$ of $q^{-1}(c)$ with smooth boundary and a map $\pi : B(c) \longrightarrow c$ making $B(c)$ a bundle over $c$. The fibre of this bundle is $T(c)$, the transverse manifold at any point of $q^{-1}(c)$. Further there is a fibre preserving map $\tilde{f} : B(c) \longrightarrow N_{\bar{f}|_c}$ such that $e \circ \tilde{f} = f \mid B(c)$.

Remark. Since $q : S(f) \cap q^{-1}(c) \longrightarrow c$ is a homeomorphism we may regard $B(c)$ as a bundle over $q^{-1}(c) \cap S(f)$ as well as over $c$.

Proof. Choose a covering of $c$ by a collection $\mathcal{I}$ of open intervals such that for each $I \in \mathcal{I}$, $e \mid p^{-1}(I)$ is an embedding. Let $\;_{I}$ : $I \times \mathbb{R} \longrightarrow p^{-1}(I)$ be a trivialization. It is no restriction to assume that for each $I$, if $J = [-1,1]$, then

$$e \circ \vartheta_I = \psi_I : I \times J \longrightarrow \mathbb{R}^2 \quad \text{satisfies:}$$

1) $\mathrm{proj} \circ \psi_I^{-1} \circ f : f^{-1}(\psi_I(I \times J)) \longrightarrow I$ is a trivial bundle (see §1.2), and

2) $\bigcup_{I \in \mathcal{I}} \vartheta_I(I \times J) = \tilde{N}_c \subseteq N_{\bar{f}}\big|_c$ is a smooth submanifold with smooth boundary $\partial \tilde{N}_c$ transverse to the fibres of $p$. $\tilde{N}_c$ is a sub-bundle of $N_{\bar{f}}\big|_c$ with fibre $J$.

As we did in §1.2, construct a CPN of $f$ for every point of $q^{-1}(I)$,

$$\begin{array}{ccc} I \times T(c) & \xrightarrow{\ \phi_I\ } & M \\ {\scriptstyle 1\times g_I}\downarrow & & \downarrow{\scriptstyle f} \\ I \times J & \xrightarrow{\ \psi_I\ } & \mathbb{R}^2 \end{array}$$

where $\phi_I$ is a diffeomorphism onto $B(c)\big|_I$, the component of $f^{-1}\psi_I(I \times J)$ containing $q^{-1}(I)$. We define $f_I$ and $\Pi_I$ by the commutativity of:

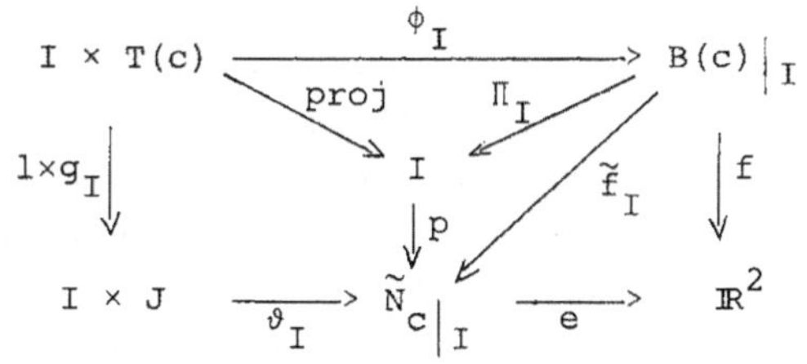


i.e. $\Pi_I = \mathrm{proj} \circ \phi_I^{-1}$, $\tilde{f}_I = (e\big|_{\tilde{N}_c|_I})^{-1} \circ f$.

Setting $B(c) = \bigcup_{I \in \mathcal{I}} B(c)\big|_I$, it is obvious by construction that the maps $\Pi_I$ cohere to define a global projection, $\Pi : B(c) \longrightarrow c$, making $B(c)$ a bundle over $c$ with fibre $T(c)$. Since $q^{-1}(I) = \phi_I((1 \times g_I)^{-1}(I \times o)) \subseteq B(c)\big|_I$ it is clear that $B(c)$ is a neighborhood of $q^{-1}(c)$. Also since $B(c)|_{I\cap I'} \subseteq (f^{-1}\circ e)(\tilde{N}_c|_{I\cap I'})$, and $\tilde{f}_I = (e|_{\tilde{N}_c|_I})^{-1}\circ f|B(c)|_{I}$, $\tilde{f}_I|B(c)|_{I\cap I'} = \tilde{f}_{I'}|B(c)|_{I\cap I'}$. Thus the $\tilde{f}_I$ cohere to give a fibre preserving map of $B(c)$ onto $\tilde{N}_c \subseteq N_{\overline{f}}$, such that $e\circ\tilde{f} = f \mid B(c)$. //

Given any bundle $\Pi : B(c) \longrightarrow c$ as in the preceding proposition, let $N(c) = q(B(c))$ and define $\pi : N(c) \longrightarrow c$ by $\Pi = \pi\circ q$. By construction $B(c) = q^{-1}(N(c))$ and $\pi : N(c) \longrightarrow c$ is a bundle with fibre $q(T(c))$.

We have $M$ as a union of three types of subsets.

1) $B(R)$, circle bundles over $R$ for $R \in \mathcal{R}$.
2) $B(c)$, transverse manifold $-T(c)$-bundles over $c$ for $c \in \mathcal{C}$
3) $B(v)$, diffeomorphic images of $I \times T(v)$ for $v \in V$, where $T(v)$ is the transverse manifold to a vertex $v$.

The three possible $T(c)$ and the four possible $T(v)$ are given in §1.3 and §1.4 respectively. $W$ is the union of the $q$-images of these sets: $R$, $N(c)$, $N(v)$ for $R \in \mathcal{R}$, $c \in \mathcal{C}$ and $v \in V$. We know something about the bundles $N(c)$. If $c \in \mathcal{C}_0$, then $N(c)$ is homeomorphic to $c \times [0,1]$ where $c$ is identified as $c \times \{0\}$. If $c \in \mathcal{C}_1$, $q(T(c))$ is homeomorphic to either a $Y$ or if $T(c)$ is not orientable to an interval $J = [-1,1]$. If $N(c)$ is a $Y$-bundle $c$ is identified as $c \times \{\text{branching point}\}$ and if $N(c)$ is a $J$-bundle then $c$ is identified as $c \times \{0\}$.

<u>Proposition</u>. <u>For $c \in \mathcal{C}_1$, the only $N(c)$ bundles that arise are trivial $Y$ and $J$ bundles and the non-trivial $Y$ bundle in which the two arms of the $Y$ are interchanged.</u>

<u>Proof</u>. Since $\overline{f} : N(c) - c \longrightarrow \mathbb{R}^2$ is an immersion, we know that $N(c) - c$ is some number or cylinders. In these cases $\overline{f}$ is an open map (see Lemma 1 of §1.3), so $\overline{f}(N(c) - c)$ must be on both sides of $c$ so $N(c) - c$ must have at least two components. Thus only the trivial $J$-bundle can arise. This eliminates two $Y$ bundles over a circle, namely those that rotate the $Y$, a third of a turn in either sense travelling around the circle. Among the three remaining non-trivial bundles the only one that can arise just interchanges the arms of the $Y$ since giving the standard orientation to $c$, the image of the stem of the $Y$ is to the right of the image of $c$, and the two arms are mapped to the left. If the part of $N(c) - c$ that contains the arms of the $Y$ is connected, the bundle is the non-trivial one mentioned. Otherwise $N(c)$ is the trivial $Y$-bundle. //

# CHAPTER II

## Canonical Coordinatized Product Neighborhoods ($C^2PN$)



### 2.0 Introduction and Summary

The object of this chapter is to choose, once and for all, coordinatized product neighborhoods for the vertices, $v \in V$, and for $c \in C$ so that we will have canonical coordinate expressions for $f$ in neighborhoods of the arcs of $S(f)$ near the points of $q^{-1}(v) \cap S(f)$ and in neighborhoods of the arcs of $S(f)$ which lie above arcs, $c \in C$. We restrict ourselves to the M-orientable case.

In §2.2 we show how to alter a given CPN of $f$ at $v$ to obtain a CPN of $f$ at $v$, $(\phi,\psi,g)$, so that $g$, in a neighborhood of $S(1\times g) \subseteq I \times T(v)$, has a canonical coordinate expression in terms of the coordinates in $\mathbb{R}^3$ (which contains $I \times T(v)$ via a standard immersion). Here the canonical coordinate expressions for $g$ are:

$$g(u,(x,y)) = \begin{cases} \{\pm(x\mp x_0)^2 - y^2 + u\}, & \text{for } v \text{ a non-simple vertex} \\ y^2 - xu + \varepsilon x^3, & \text{for } v \text{ a cusp} \end{cases}$$

In §2.3, we generalize the notion of CPN to all of a $q$-saturated neighborhood $B(c)$ of $c$ to a commutative diagram

$$\begin{array}{ccc} I \times T(c) & \xrightarrow{\ \phi\ } & B(c) \\ {\scriptstyle 1 \times g}\downarrow & & \downarrow{\scriptstyle f} \\ I \times J & \xrightarrow{\ \psi\ } & N_c \end{array}$$

where $\phi$ and $\psi$ are diffeomorphisms. In Proposition 2 we show that if $c$ joins vertex $v_-$ to $v_+$, that if $I_-$ and $I_+$ are initial and terminal segments of $I$ that there is a CPN of $f$ along $c$, $(\phi,\psi,g)$, so that in the diagram below, the restrictions of the maps to neighborhoods of $I_\pm \times 0$ in $I_\pm \times T(c)$ and $I_\pm \times J$ are canonically determined. A complete catalog of canonical forms is given as part of Proposition 2.

$$\begin{array}{ccc} I_\pm \times T(c) & \xrightarrow{\phi_{v_\pm}^{-1}\circ\phi} & I \times T(v) \\ {\scriptstyle 1\times g}\downarrow & & \downarrow{\scriptstyle \tilde{f}} \\ I_\pm \times J & \xrightarrow{\psi_{v_\pm}^{-1}\circ\psi} & \tilde{N}_c \end{array}$$

Finally the theorem of §2.3 shows that the CPN of $f$ along $c$ can be chosen so that, in addition, in a neighborhood of $I \times 0$ in $I \times T(c)$, $g(u,x,y) = x^2 + (-1)^j y^2$ if $c \in C_j$, $j = 0, 1$. An analogous result is proven for $c$, a closed element of $C$.

Lemma 0 of 2.1 is the principal tool used to modify the maps of CPN's. It allows us to alter a given map on an open subset $S$ of its domain in such a way that in a slightly smaller subset $S' \subseteq S$, the given map has been replaced by one equivalent to it. In part, this lemma describes a situation in which a diffeomorphism on a subset $D$ of $\mathbb{R}^N$, can be changed in the neighborhood of the boundary of $D$ so as to be extendable by the identity map to a diffeomorphism of all of $\mathbb{R}^N$.

## 2.1 A technical lemma and its corollaries

Notation. If $S_0$, $S_1$ are subsets of a topological space we say $S_0$ is nested in $S_1$ or $S_0 \subseteq S_1$ are nested if $\bar{S}_0 \subseteq S_1^0$.

By an interval $I$ in $\mathbb{R}^m$ we mean the $m$-fold product of real intervals parallel to the coordinate axes with $0$ in their interiors. An interval $I_1$, in $\mathbb{R}^m$ is a subinterval of $I$ if $I_1$ is nested in $I$.

Remark. In the following lemma, $T$ is an interval in $\mathbb{R}^p$, and the metric space in which everything is defined is $T \times \mathbb{R}^q \times \mathbb{R}^r \times \mathbb{R}^n$; all topological notions refer to this space and not to the containing $\mathbb{R}^m$, $m = p + q + r + n$.

Lemma 0. Let $T$ be an interval of $\mathbb{R}^p$, and $J$ and $K$ be compact intervals of $\mathbb{R}^q$ and $\mathbb{R}^r$ respectively. Let $N$ be a closed neighborhood of $T \times J \times K \times \{0\}$ in $T \times J \times K \times \mathbb{R}^n$, and let $f : N \longrightarrow \mathbb{R}^n$ define an orientation preserving diffeomorphism, $1 \times f : N \longrightarrow T \times J \times K \times \mathbb{R}^n$ such that

a) $f \mid T \times J \times K \times \{0\} = 0$

b) For $J_0$ a closed subinterval of $J$, $f \mid N \cap (T\times J - J_0 \times K \times \mathbb{R}^n) = \text{proj}$, projection onto the $\mathbb{R}^n$ factor.

c) For some $(t_0, v_0) \in T \times K$, $\text{Jac}_x f(t_0, \cdot, v_0, 0) : J \longrightarrow GL(n,\mathbb{R})$ is homotopic to the map constantly the identity, $I \in GL(n,\mathbb{R})$ such that the homotopy is constantly $I$ on $J - J_0$. (Here $\text{Jac}_x f(t_0,u,v_0,0)$ is the Jacobian matrix of $f \mid (\{t_0,u,v_0\} \times \mathbb{R}^n) \cap N$ evaluated at $0 \in \mathbb{R}^n$), then for any closed subintervals $J_0 \subseteq J_1 \subseteq J$ and $K_0 \subseteq K_1 \subseteq K$ and sufficiently small closed nested neighborhoods of $T \times J \times K \times \{0\}$, $N_0 \subseteq N_1 \subseteq N$ there are orientation preserving diffeomorphisms: $1 \times g : T \times \mathbb{R}^q \times \mathbb{R}^r \times \mathbb{R}^n \longrightarrow T \times \mathbb{R}^q \times \mathbb{R}^r \times \mathbb{R}^n \longleftarrow N : 1 \times h$ such that $g \mid T \times \mathbb{R}^q \times \mathbb{R}^r \times \{0\} = 0 = h \mid T \times J \times K \times \{0\}$, and

1) On $N_0 \cap (T\times J\times K_0\times\mathbb{R}^n)$, $g = f$, $h = \text{proj}$.

2) On $T \times \mathbb{R}^q \times \mathbb{R}^r \times \mathbb{R}^n - N_1 \cap (T\times J_1\times K_1\times\mathbb{R}^n)$, $g = \text{proj}$ and on $N - N_1' \cap (T\times J_1\times K_1\times\mathbb{R}^n)$, $h = f$, where $N_1' = (1\times f)^{-1}(N_1)$.

Remark. If $T$ is also assumed compact, the neighborhoods, $N_0$ and $N_1$ may be taken in the form $T \times J \times K \times B_i$ where $B_0 \subseteq B_1$ as nested balls in $\mathbb{R}^n$.

We state two special cases that are used a number of times in the sequel.

If $q = r = 0$:

Corollary 1. Let $T$ be an interval in $\mathbb{R}^p$. Let $N$ be a neighborhood of $T \times \{0\} \subseteq T \times \mathbb{R}^n$. Let $f : N \longrightarrow \mathbb{R}^n$ define an orientation preserving diffeomorphism $1 \times f : N \longrightarrow T \times \mathbb{R}^n$ such that $f \mid T \times \{0\} = 0$, then for sufficiently small closed nested neighborhoods of $T \times \{0\}$, $N_0 \subseteq N_1 \subseteq N$ there is an orientation preserving diffeomorphism $1 \times g$ of $T \times \mathbb{R}^n$ such that $g \mid T \times \{0\} = 0$, and

(a) On $N_0$, $g = f$

(b) On $(T\times\mathbb{R}^n) - N_1$, $g = \text{proj}$.

If $p = q = 0$:

Corollary 2. Let $K$ be a compact interval of $\mathbb{R}^r$, and $B$ a closed ball about $0$ in $\mathbb{R}^n$. Let $1 \times f : K \times B \longrightarrow K \times \mathbb{R}^n$ be an *orientation preserving diffeomorphism with* $f \mid K \times \{0\} = 0$. *Then for sufficiently small closed nested balls,* $B_0 \subseteq B_1 \subseteq B \subseteq \mathbb{R}^n$ *and any closed nested intervals* $K_0 \subseteq K_1 \subseteq K \subseteq \mathbb{R}^r$ *there is an orientation preserving diffeomorphism* $1 \times g : \mathbb{R}^r \times \mathbb{R}^n \longrightarrow \mathbb{R}^r \times \mathbb{R}^n$ *such that* $g \mid \mathbb{R}^r \times \{0\} = 0$ *and:*

(a) *On* $K_0 \times B_0$, $g = f$

(b) On $\mathbb{R}^n \times \mathbb{R}^r - (K_1 \times B_1)$, $g = \text{proj}$.

The following is a corollary of the proof and not a formal consequence of the statement of the lemma itself.

*Corollary 3. Let* $T_0 \subseteq T_1 \subseteq T$ *be a nest of intervals in* $T \subseteq \mathbb{R}^p$ *and let* $N$ *be a neighborhood of* $T \times \{0\}$ *in* $T \times \mathbb{R}^n$. *Let* $f : N \longrightarrow \mathbb{R}^n$ *define an orientation preserving diffeomorphism* $1 \times f : N \longrightarrow T \times \mathbb{R}^n$ *such that* $f \mid T \times \{0\} = 0$. *Then for sufficiently small neighborhoods of* $N_0$ of $T_0 \times \{0\}$ *and* $N_1$ of $T_1 \times \{0\}$, *there is an orientation preserving diffeomorphism* $1 \times g$ *of* $T \times \mathbb{R}^n$ *such that* $g \mid T \times \{0\} = 0$ *and*

(a) *On* $N_0$, $g = f$

(b) *On* $T \times \mathbb{R}^n - N_1$, $g = \text{proj}$.

To get this corollary the proof must be modified by making bump functions depend on $t \in T$ as well as on $x \in \mathbb{R}^n$.

To prove Lemma 0 it suffices to consider two cases: 1) in which $\text{Jac}_x f(\omega,0) = I$, for all $\omega \in T \times J \times K$, and $f(\omega,x) = A(\omega)x$ where $A : T \times J \times K \longrightarrow GL(n,\mathbb{R})$. Case 2 is divided into three parts by writing $A = mQO$, where $m \geqslant 1$, $Q$ is positive definite with all its eigenvalues $\leqslant 1$, and $O$ is orthogonal. We relegate the proof of this lemma to the appendix.

## 2.2 *Canonical coordinatized product neighborhoods of* $f$ *near the vertices*

*Notation.* Here and in the sequel we will use the following notations. For two dimensional surfaces: $\mathbb{D}$, disc; $\mathbb{S}$, sphere; $\mathbb{T}$, torus. If $\mathbb{A}$ is any two dimensional surface, $\mathbb{A}_{(k)}$ is $\mathbb{A}$ with $k$ discs deleted.

In this paragraph we return to consideration of our map $f : M \longrightarrow \mathbb{R}^2$, $f \in S$. We restrict our attention from now on to the case of $M$ an *orientable* compact 3-manifold.

Let $v$ be a vertex of $W$; that is $v$ is either a cusp or a non-simple point of $W$. There are three different types of vertices. The transverse manifolds $T(v)$ for each of these are: $\mathbb{D}_{(1)}$ if $v$ is a cusp, $T_{(2)}$ if $v$ is of type $(1\cdot 2\cdot 2\cdot 1)$, $\mathbb{D}_{(3)}$ if $v$ is of type $(1\cdot 2\cdot 2\cdot 3)$. In this paragraph we show that there are (CPN)'s, coordinatized product neighborhoods, $(\phi,\psi,g)$ such that near $\phi^{-1}(q^{-1}(v)\cap S(f))$, the maps $g$ have canonical forms.

For each $v$ we choose one such $(\phi_v,\psi_v,g_v)$ once and for all and call such coordinatized product neighborhoods, *canonical*, ($C^2PN$). We orient $I\times T(v)$ and $I\times J$ so that the $\phi_v$, $\psi_v$ are orientation preserving. We will specify, somewhat arbitrarily, an orientation in $I$ (the same orientation in both $I\times T(v)$ and $I\times J$), which will determine the orientation of $T(v)$ and $J$.

We first prove a general lemma that shows that any CPN for $f$ at $v$ can be modified so that the resulting map $g$ has a canonical form in a neighborhood of each point of $\phi^{-1}(q^{-1}(v)\cap S(f))$. The rest of this paragraph is essentially a tabulation of the application of this lemma to (CPN)'s for the three types of vertices.

*Lemma. For any* $w\in W$ *let* $(\phi,\psi,g)$ *be a* CPN *of* $f$ *at* $w$

$$\begin{array}{ccc} I\times T & \xrightarrow{\ \phi\ } & M \\ {\scriptstyle 1\times g}\downarrow & & \downarrow{\scriptstyle f} \\ I\times J & \xrightarrow{\ \psi\ } & \mathbb{R}^2 \end{array}$$

*Let* $(0,P)\in\phi^{-1}(q^{-1}(w))$. *Suppose that the germ of* $1\times g$ *at* $(0,P)$ *is equivalent via orientation preserving diffeomorphisms to a germ* $1\times k$ *at* $(0,P)$. *Then there is a* CPN *of* $f$ *at* $w$

$$\begin{array}{ccc} I'\times T & \xrightarrow{\ \phi'\ } & M \\ {\scriptstyle 1\times g'}\downarrow & & \downarrow{\scriptstyle f} \\ I'\times J' & \xrightarrow{\ \psi'\ } & \mathbb{R}^2 \end{array}$$

*such that for discs,* $\mathbb{D}_0\subseteq\mathbb{D}_1\subseteq T$, *about* $P$, *we have*

$$\begin{cases} g'=k & \text{on } I'\times\mathbb{D}_0 \\ g'=g & \text{on } I'\times(T-\mathbb{D}_1). \end{cases}$$

Proof. This is an application of Lemma 0 of §2.1. Since the size of $I'$ and $J'$ in the CPN is irrelevant--we can always change scale--we drop the primes. Suppose the equivalence between $1 \times g$ and $1 \times k$ is given by diffeomorphisms $\alpha$ and $\beta$:

$$\begin{array}{ccc} I \times \mathbb{D} & \xrightarrow{\alpha} & I \times \mathbb{D} \\ {\scriptstyle 1 \times k}\downarrow & & \downarrow{\scriptstyle 1 \times g} \\ I \times J & \xrightarrow{\beta} & I \times J \end{array}$$

Notice that $(\psi\circ\beta) \mid (0\times J) \pitchfork f$, for if $\psi\circ\beta(0,y) = f(z)$

$$(\psi\circ\beta)_\star T(0\times J)_{(0,y)} + f_\star(TM)_z = T\mathbb{R}^2_{f(z)} \quad \text{iff}$$

$$T(0\times J)_{(0,y)} + ((\psi\circ\beta)^{-1}\circ f)_\star TM_z = T\mathbb{R}^2_{(0,y)}.$$

But $((\psi\circ\beta)^{-1}\circ f)_\star TM_z = (1\times k\circ\alpha^{-1}\circ\phi^{-1})_\star TM_x = (1\times k)_\star T(I\times\mathbb{D})_{(0,x)}$, where $\phi\circ\alpha(0,x) = z$, and obviously $(1\times k_\star)T(I\times\mathbb{D})_{(0,x)} + T(0\times J)_{(0,y)} = T\mathbb{R}^2_{(0,y)}$.

By cutting down the size of $I$, $J$, and $\mathbb{D}$ if necessary we can replace the map $\psi$ in the construction of the CPN by $\psi' = \psi\circ\beta$ and obtain a new CPN as part of the following diagram:

$$\begin{array}{ccccc} I \times \mathbb{D} & \xrightarrow{1 \times \vartheta} & I \times T & \xrightarrow{\tilde{\phi}} & M \\ & {\scriptstyle 1 \times k}\searrow & \downarrow{\scriptstyle 1 \times \tilde{g}} & & \downarrow{\scriptstyle f} \\ & & I \times J & \xrightarrow{\psi'} & \mathbb{R}^2 \end{array}$$

where $1 \times \vartheta = \tilde{\phi}^{-1}\circ\phi_V\circ\alpha$ (see §1.2 for the definitions of $\tilde{\phi}$ and $\tilde{g}$). The square commutes by construction. To see that the triangle commutes: $(1\times\tilde{g})\circ(1\times\vartheta) = 1 \times k$, it is enough to show $\psi'\circ(1\times\tilde{g})\circ(1\times\vartheta) = \psi'\circ(1\times k)$. But $\psi'\circ(1\times\tilde{g})\circ(1\times\vartheta) = f\circ\tilde{\phi}\circ(1\times\vartheta) = f\circ\phi_V\circ\alpha = \psi\circ(1\times g)\circ\alpha = \psi\circ\beta\circ(1\times k) = \psi'\circ(1\times k)$.

Now by Lemma 0 of §2.1 there are nested discs $\mathbb{D}_0 \subseteq \mathbb{D}_1 \subseteq \mathbb{D}$ about $P$ and a diffeomorphism $1 \times \vartheta : I \times T \circlearrowleft$ which agrees with $1 \times \vartheta$ on $I \times \mathbb{D}_0$ and which is the identity map on $I \times T - \mathbb{D}_1$. The CPN that we're after is:

$$\begin{array}{ccc} I \times T & \xrightarrow{\tilde{\phi}\circ(1\times\vartheta') \;=\; \phi'} & M \\ \Big\downarrow 1\times g' & & \Big\downarrow f \\ I \times J & \xrightarrow{\;\;\psi'\;\;} & \mathbb{R}^2 \end{array}$$

where $g'$ is defined by: $\psi'\circ(1\times g') = f\circ\phi'$. // Lemma.

In the following propositions we will apply this lemma and the well known canonical forms for the germs at a cusp or a fold point [Mo].

(i) If $v$ is a cusp, one of the components of $C$, say $c$, that abuts $v$ is in $C_0$. The composition of $\phi_v^{-1} : c \cap B(v) \longrightarrow I \times T(v)$ and $\text{proj} : I \times T(v) \longrightarrow I$ is an embedding onto an interval of $I - \{0\}$. We orient $I$ so that the embedding is orientation preserving. (We take the standard orientation of $c$ (see §1.5).) In this case $T(v) = \$_{(2)} - \mathbb{R}^2$ as pictured below with orientation $dx \wedge dy$; we use this embedding to give coordinates on $\$_{(2)}$. If the coordinates of $I$, $\$_{(2)}$ are $u$, $(x,y)$ respectively then $du$ orients $I$ and $du \wedge dx \wedge dy$ orients $I \times \$_{(2)}$.

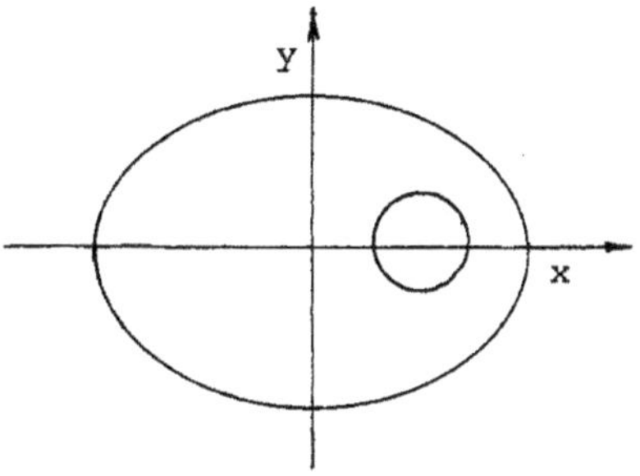


$\$_{(2)}$

Proposition 1 (cusp). For any cusp, $v$, of the mapping, $f$, there is a coordinatized product neighborhood

$$\begin{array}{ccc} I \times S_{(2)} & \xrightarrow{\;\;\phi_v\;\;} & B(v) \subseteq M \\ \Big\downarrow 1\times g_v & & \Big\downarrow f \\ I \times J & \xrightarrow{\;\;\psi_v\;\;} & \mathbb{R}^2 \end{array}$$

with the cusp at $\phi_v(0,0)$, such that for a sufficiently small disc $\mathbb{D} \subseteq \$_{(2)}$ about $0$,

$$g_v \mid I \times \mathbb{D}(u,x,y) = y^2 - xu + \varepsilon x^3, \qquad (\varepsilon = \pm 1).$$

We sketch $I \times \$_{(2)}$ and $I \times J$:

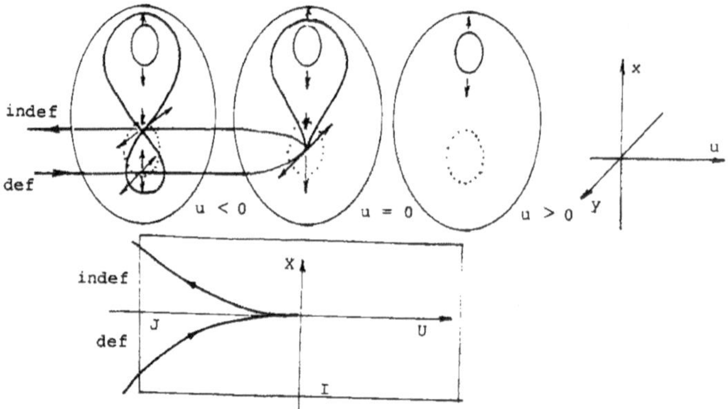


$$g_V(u,x,y) = y^2 - xu - x^3$$
$$(\varepsilon = -1)$$

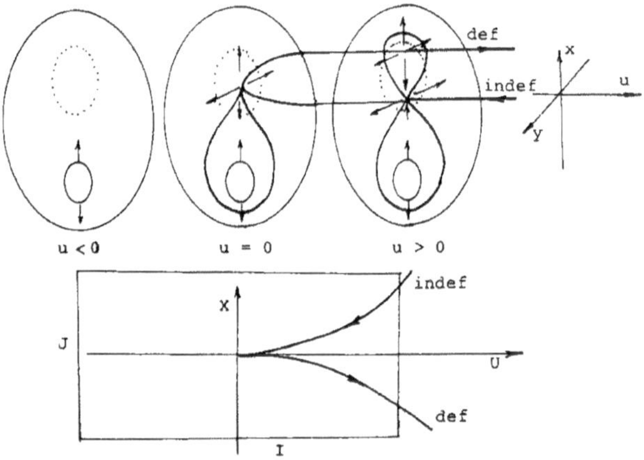


$$g_V(u,x,y) = y^2 - xu + x^3$$
$$(\varepsilon = +1)$$

In the figures of $u \times \$_{(2)}$, we've indicated the position of $u \times \mathbb{D}$ with a dotted circle. The dark curves are the f-preimages of the singular points of $f$ pulled back by $\phi_V$ and the light arrows indicate the directions of increasing $g_V$-values. In the $I \times J$ figures, the dark labelled curves are the $f$-images of $S(f)$ pulled back by $\psi_V$. The arcs of the $\phi_V$ and $\psi_V$ preimages of $S(f)$ and $f(S(f))$ are given the standard orientations. In the analogous figures that follow we will follow the same conventions.

(ii) If $v$ is a $(1\cdot2\cdot2\cdot1)$ point, the transverse manifold is a torus with two discs removed, $\mathbb{T}_{(2)}$. We immerse $\mathbb{T}_{(2)}$ in $\mathbb{R}^2$ as shown. In the description given in the following proposition the coordinates on the discs around $P_\pm$ are those inherited from $\mathbb{R}^2$ via this immersion. The coordinates of $P_\pm$ are $(\pm x_0, 0)$.

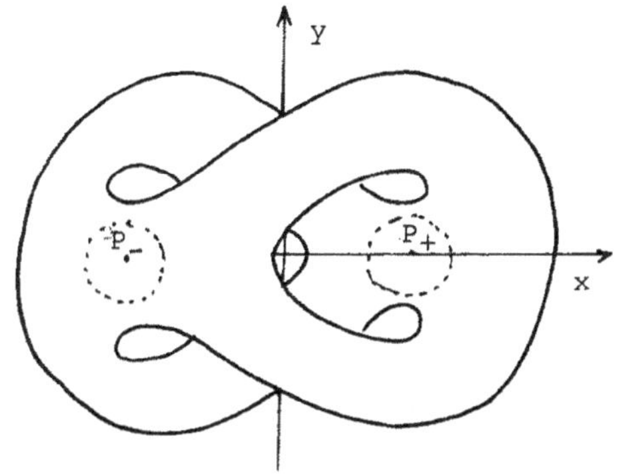


Proposition 1 $(1\cdot2\cdot2\cdot1)$. For any $(1\cdot2\cdot2\cdot1)$-point $v$ of the mapping $f$, there is a coordinatized product neighborhood of $f$ at $q^{-1}(v) \cap S(f)$

$$\begin{array}{ccc} I \times \mathbb{T}_{(2)} & \xrightarrow{\ \phi_v\ } & B(v) \subseteq M \\ \Big\downarrow {\scriptstyle 1 \times g_v} & & \Big\downarrow {\scriptstyle f} \\ I \times J & \xrightarrow{\ \psi_v\ } & \mathbb{R}^2 \end{array}$$

with $\phi_v^{-1}(S(f) \cap B(v)) = I \times \{P_-, P_+\}$ and $\phi_v^{-1}(S(f) \cap q^{-1}(v)) = 0 \times \{P_-, P_+\}$, such that if $\mathbb{D}_\pm$ are small enough discs around $P_\pm$ in $\mathbb{T}_{(2)}$, then

$$g_v \mid I \times \mathbb{D}_\pm(u,x,y) = \pm\{(x \mp x_0)^2 - y^2 + u\}.$$

Note. It is easy to check that it is no restriction to assume that $g_v \mid I \times \mathbb{D}$ has the prescribed form since by orientation preserving coordinate changes it can be transformed into any of the forms:

$$\pm\{\alpha[(x \mp x_0)^2 - y^2] + \beta u\}, \quad \alpha = \pm 1, \quad \beta = \pm 1.$$

This coordinatized product neighborhood is not uniquely determined on $I \times \mathbb{D}_\pm$ by $g_v \mid I \times \{P_-, P_+\}$ since $g_v(u, \pm x_0, 0) = \pm u$, and using $\phi'_v(u,x,y) = \phi_v(-u,x,-y)$ and $\psi'_v(U,X) = \psi_v(-U,-X)$ the resulting $g'_v(u, \pm x_0, 0) = u$. However it is uniquely determined if, in addition to $g_v(u, \pm x_0, 0) = \pm u$, we specify on which component of $I - \{0\} \times P_+$,

$\phi_V$ is an orientation preserving embedding into the component of $S(f) - V$. The coordinate map $\phi_V$ is orientation preserving on one of these components and $\phi_V'$ is orientation reversing on the other. In the sequel, we assume we have arbitrarily chosen one of the components of $(I-\{0\}) \times P_+$ on which $\phi_V$ is orientation preserving therewith fixing a coordinatized product neighborhood as given by the proposition.

We sketch $I \times \mathbb{T}_{(2)}$ and $I \times J$.

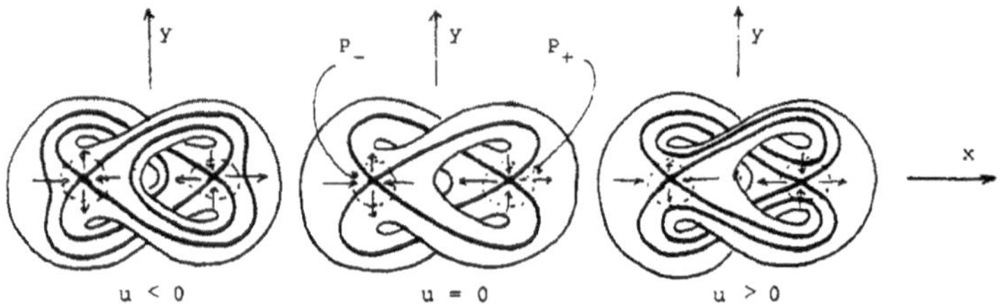


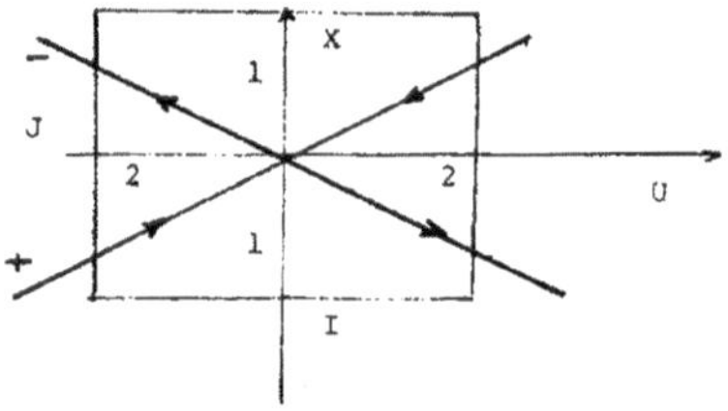


(iii) If $v$ is a $(1{\cdot}2{\cdot}2{\cdot}3)$-point, the transverse manifold is $\$_{(4)}$, a disc with three holes. We take a standard embedding of $\$_{(4)}$ in $\mathbb{R}^2$ as shown below. In the proposition below we will use coordinates on the discs about $P_\pm$ given by this embedding. The coordinates of $P_\pm$ are $(\pm x_0, 0)$. We choose an orientation of $I$ so that $\phi_V \mid I - \{0\} \times \{P_-, P_+\}$ is an orientation preserving map into the oriented components of $C_1$, and as usual $du$ orients $I$ and $du \wedge dx \wedge dy$ orients $I \times \$_{(4)}$.

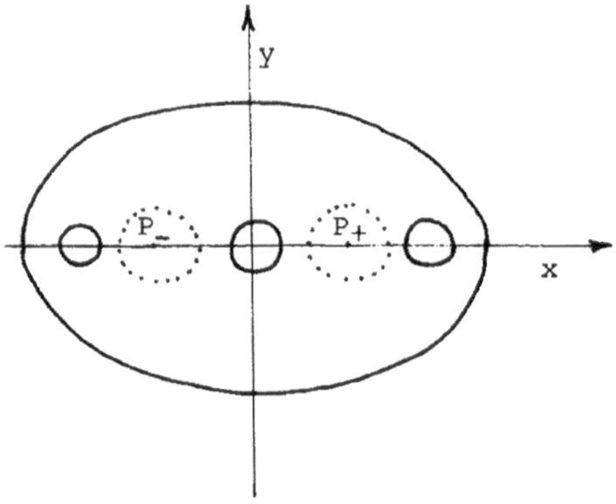


Proposition 1 $(1{\cdot}2{\cdot}2{\cdot}3)$. *For any $(1{\cdot}2{\cdot}2{\cdot}3)$-point $v$ of the mapping $f$, there is a coordinatized product neighborhood of $f$ at $q^{-1}(v) \cap S(f)$*

$$\begin{array}{ccc} I \times \$_{(4)} & \xrightarrow{\phi_v} & B(v) \subseteq M \\ \downarrow 1 \times g_v & & \downarrow f \\ I \times J & \xrightarrow{\psi_v} & \mathbb{R}^2 \end{array}$$

<u>with</u> $\phi_v^{-1}(S(f) \cap B(v)) = I \times \{P_-, P_+\}$ <u>and</u> $\phi_v^{-1}(S(f) \cap q^{-1}(v)) = \{0\} \times \{P_-, P_+\}$, <u>such that for sufficiently small discs</u> $\mathbb{D}_\pm$ <u>about</u> $P_\pm$ <u>in</u> $\$_{(4)}$,

$$g_v \mid I \times \mathbb{D}_\pm(u,x,0) = (x \mp x_0)^2 - y^2 \pm u.$$

We sketch $I \times \$_{(4)}$ and $I \times J$.

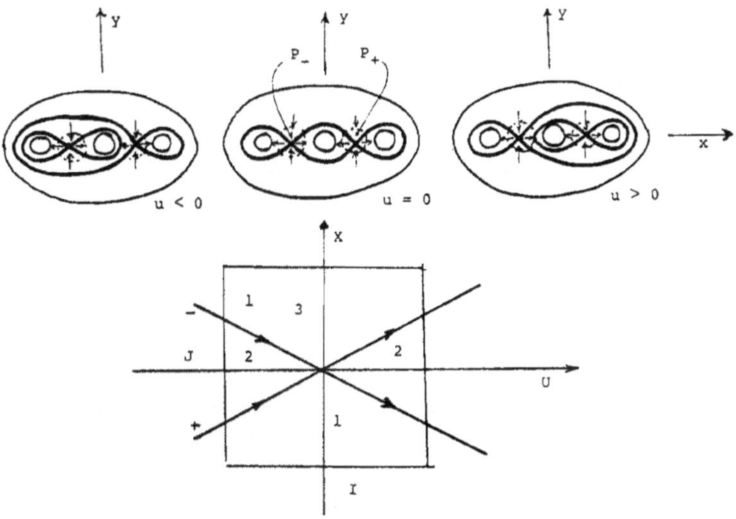


<u>Note</u>. It is again easy to check that it is no restriction to assume that $g_v \mid I \times \mathbb{D}$ has the prescribed form since by orientation preserving coordinate changes it can be transformed into any of the forms

$$\alpha[(x \mp x_0)^2 - y^2] \pm \beta u, \quad \alpha = \pm 1, \quad \beta = \pm 1.$$

Henceforth our coordinatized product neighborhoods of $f$ at $q^{-1}(v) \cap S(f)$ will be taken as given by the three parts of <u>Proposition 1</u>, the mappings $g_v$ having the given canonical forms on the neighborhoods of the singular curves. In the case of $v$ a $(1 \cdot 2 \cdot 2 \cdot 1)$-point, the lack of uniqueness in the description of a coordinatized product neighborhood has been eliminated by an arbitrary choice of one of two possibilities. For each vertex $v \in V$, our fixed CPN, $(\phi_v, \psi_v, g_v)$ as described in Proposition 1 is called the canonical coordinatized product neighborhood of $f$ at $v$, ($C^2PN$ of $f$ at $v$).

## 2.3 Canonical coordinatized product neighborhoods of $f$ along arcs of $\mathcal{C}$

In this paragraph we generalize the notion of CPN of $f$ at a point to that of CPN of $f$ along a component $c \in \mathcal{C}$. We introduce canonical forms into these coordinatizations of the neighborhoods $B(c)$ of $q^{-1}(c)$ for $c \in \mathcal{C}$ (see §1.6) in tubular neighborhoods of $q^{-1}(C) \cap S(f)$ analogous to the canonical forms introduced in the preceding paragraph in a tubular neighborhood of $S(f) \cap B(v)$. Furthermore the canonical coordinatizations in $B(c)$ are chosen so that if $c$ abuts a vertex $v$, then the two coordinatizations of the neighborhood of $S(f) \cap B(c) \cap B(v)$ are related in a canonical way.

Recall (§1.6) that in a neighborhood $B(c)$ of $q^{-1}(c)$, $c \in \mathcal{C}$, we factor $f$ as:

$$B(c) \xrightarrow{\tilde{f}} \tilde{N}_c \overset{e}{\hookrightarrow} \mathbb{R}^2$$

where $B(c)$ and $\tilde{N}_c$ are bundles over $c$. The fibre of $B(c)$ is $T(c)$, a transverse manifold at any point of $q^{-1}(c) \cap S(f)$, and the fibre of $\tilde{N}_c$, a sub bundle of the normal bundle to the immersion $\bar{f}|_c$, is a compact interval, $J \subseteq \mathbb{R}$. For $c \in \mathcal{C}_0$, $T(c) = \mathbb{D}$ and if $c \in \mathcal{C}_1$, $T(c) = \$_{(3)} = \mathbb{D}_{(2)}$ a disc with two holes.

In order to generalize the notion of a CPN of $f$ at a point to that of a CPN of $f$ along an element $c \in \mathcal{C}$, we first choose an orientation preserving immersion $\sigma : I \longrightarrow c$, $I = (-1,+1)$, which is a diffeomorphism if $c$ joins a vertex, $v_-$, to a vertex, $v_+$ or if $c$ is closed, wraps $I$ positively around $c$ with some overlap at the ends. Pulling the bundles $B(c)$ and $N_c$ back over $I$ by $\sigma$ gives trivial bundles. If we choose trivializations for these bundles we obtain a commutative diagram

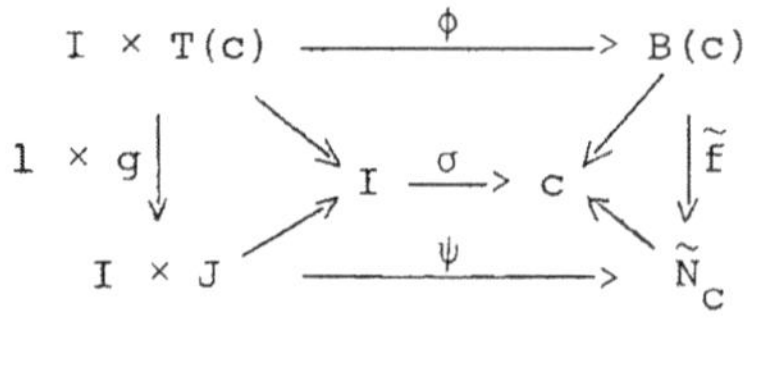

where $T(c)$ and $J$ are oriented so that $\phi$ and $\psi$ are orientation preserving local diffeomorphisms (bundle morphisms over $\sigma$). We call such a triple of mappings $(\phi,\psi,g)$ a <u>coordinatized product neighborhood of f along c</u> (CPN) of $f$ along $c$).

The first step toward producing what we will call a canonical CPN of $f$ along $c$ uses our choices of $(\phi_v,\psi_v,g_v)$, $C^2$PN of $f$ at $v \in V$. We may assume that our immersion $e$ of $\tilde{N}_c$ in $\mathbb{R}^2$ is an embedding of the part of $\tilde{N}_c$ above $c \cap N(v)$ for any $v$ that $c$ abuts and that $e(\tilde{N}_c|_{c\cap N(v)}) \subseteq \psi_v(I\times J)$. Further we choose $e$ so that $\psi_v^{-1}\circ e : \tilde{N}_c|_{c\cap N(v)} \longrightarrow I \times J$ is a fibre preserving embedding. It is clear that the composition $\mathrm{proj}\circ\psi_v^{-1}\circ e\circ(\text{zero section}) : c \cap N(v) \longrightarrow I$ embeds $c \cap N(v)$ onto either $(-1,0) = I_-(1)$, $(0,1) = I_+(1)$.

From now on we assume that we've chosen the immersions $e$ of $\tilde{N}_c$ for all $c$ as described. We call these choices of $e$, <u>canonical</u>. Suppose now that $(\phi,\psi,g)$ is any CPN of $f$ along $c$ and suppose $c$ abuts a vertex, $v$. Obviously $\sigma^{-1}(c \cap N(v)) = I_i(\alpha)$ for some $1 > \alpha > 0$ where $I_i(\alpha) = \{t \in \mathbb{R} \mid 1 - \alpha < it < 1\}$, $i = \pm$. Since $e$ is canonical, in the following commutative diagram

$$\begin{array}{ccccccc}
I_i(\alpha)\times T(c) & \xrightarrow{\phi} & B(c)|_{c\cap N(v)} & \subseteq & B(v) & \xleftarrow{\phi_v} & I\times T(v)\\
\big\downarrow{\scriptstyle 1\times g} & & \big\downarrow{\scriptstyle \tilde{f}} & & \big\downarrow{\scriptstyle f} & & \big\downarrow{\scriptstyle 1\times g_v}\\
I_i(\alpha)\times J & \xrightarrow{\psi} & \tilde{N}_c|_{c\cap N(v)} & \xrightarrow{e} & \psi_v(I\times J) & \xleftarrow{\psi_v} & I\times T
\end{array}$$

the maps $\phi_v^{-1}\circ\phi$ and $\psi_v^{-1}\circ e\circ\psi$ have the form $K \times \beta$ and $K \times \gamma$ where $K : I_i(\alpha) \longrightarrow I$, is a diffeomorphism of $I_i(\alpha)$ onto one of $I_\pm(1) \subseteq I$.

Also if $c \in C$ is a closed component, we may assume that for some $\alpha > 0$, $\sigma(I_+(\alpha)) = \sigma(I_-(\alpha))$. In the diagram

$$\begin{array}{ccccc}
I_+(\alpha)\times T(c) & \xrightarrow{\phi} & B(c) & \xleftarrow{\phi} & I_-(\alpha)\times T(c)\\
\big\downarrow{\scriptstyle 1\times g} & & \big\downarrow{\scriptstyle \tilde{f}} & & \big\downarrow{\scriptstyle 1\times g}\\
I_+(\alpha)\times J & \xrightarrow{\psi} & \tilde{N}_c & \xleftarrow{\psi} & I_-(\alpha)\times J
\end{array}$$

if we let $\phi_i$, $g_i$ (and $\psi_i$) denote $\phi$, $g$, (and $\psi$) restricted to $I_i(\alpha) \times T(c)$ (and to $I_i(\alpha) \times J$), then $\phi_-^{-1}\circ\phi_+$ and $\psi_-^{-1}\circ\psi_+$ also have the form $K \times \beta$ and $K \times \gamma$ respectively where $K : I_+(\alpha) \longrightarrow I_-(\alpha)$ is an orientation preserving diffeomorphism.

For $c$ either closed or open we can always choose the map $\sigma : I \longrightarrow c$ in such a way that the maps $K$ which arise are linear. In the case of $c$ closed we just require that $\sigma(u) = \sigma(u+\alpha-2)$ for $u \in I_+(\alpha)$. For open $c$ the choice of $\sigma$ depends on the CPN's of $f$ at the vertices $c$ abuts. Call such a choice of $\sigma$, canonical. If we let $L(\alpha) : I_+(\alpha) \longrightarrow I_-(\alpha) : u \longrightarrow u + \alpha - 2$, and $L_\pm(\alpha) : I_\pm(\alpha) \longrightarrow I_\mp(1) : u \longrightarrow \frac{1}{\alpha}(u \mp 1)$, then having chosen $\sigma$ as well as $e$ canonically the maps, $K$, that arise for open $c$ are $\pm L_i(\alpha)$, $i = +,-$ and for closed $c$, $K = L(\alpha)$. (To simplify the notation a bit, we write $I_i$ for $I_i(\alpha)$, and $L$ and $L_i$ for $L(\alpha)$ and $L_i(\alpha)$.) In all that follows we will assume that all the immersions $e$ and $\sigma$ have been chosen canonically. Thus our CPN's of $f$ along $c \in C$ are somewhat adjusted to the $C^2$PN's, of $f$ at $v \in V$.

In order to describe our maps explicitly we take standard embeddings of $T(c)$ in $\mathbb{R}^2$. If $c \in C_0$, $T(c) = \mathbb{D}$, the unit disc about the origin and if $c \in C_1$, $T(c) = \mathbb{D}_{(2)}$ which we embed as shown symmetrically inside the unit disc.

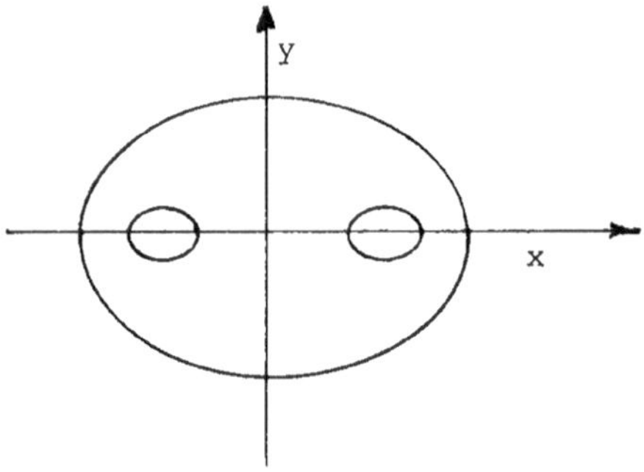


For $c$ closed, for any CPN of $f$ along $c$ we have the indicated maps in the commutative square:

$$(*_{\text{closed } c})\qquad \begin{array}{ccc} I_+ \times T(c) & \xrightarrow{\ \phi_-^{-1}\circ\phi_+ \,=\, L \times \beta\ } & I_- \times T(c) \\ {\scriptstyle 1 \times g_+}\big\downarrow & & \big\downarrow{\scriptstyle 1 \times g_-} \\ I_+ \times J & \xrightarrow{\ \psi_-^{-1}\circ\psi_+ \,=\, L \times \gamma\ } & I_- \times J \end{array}$$

where $\beta : T(c) \longrightarrow T(c) : \xi \longrightarrow \xi$ except if $c \in C_1$ and $B(c)$ is non-trivial in which case $\beta(\xi) = -\xi$, $\gamma : J \longrightarrow J : x \longrightarrow x$.

Proposition 2 (for $c$ closed). Let $c \in C_j$, $j = 0, 1$. Then there is a CPN of $f$ along $c$ such that for $(u,(x,y))$ in a small neighborhood of $I_i \times 0$ in $I_i \times T(c)$,

$$g_i(u,(x,y)) = x^2 + (-1)^j y^2, \quad i = \pm.$$

Proof. This is a straightforward application of Lemma 0 and Corollary 2 of §2.1 using the normal form for germs at fold points of $f$ [G-du P-W, Theorem 2.1 of G-W]. //

Our final objective is to construct a CPN of $f$ along $c$, $(\phi,\psi,g)$, so that $\phi_-^{-1}\circ\phi_+ = L \times \beta$ and $\psi_-^{-1}\circ\psi_+ = (L\times 1)$ and $g(u,(x,y)) = x^2 + (-1)^j y^2$ for a neighborhood of all of $I \times \{0\}$ in $I \times T(c)$ not merely on a neighborhood of $I_\pm \times \{0\}$ which we now have. Before getting this "global" canonical form, we prove the version of Proposition 2 for nonclosed $c \in C$.

Now suppose $c \in C$ is open and joins vertex $v_-$ to vertex $v_+$. Suppose that $q^{-1}(c) \cap S(f)$ joins $v_-^*$ to $v_+^*$ where $v_i^* \in q^{-1}(v_i) \cap S(f) \subseteq T(v_i)$ and let $\mathbb{D}_i^* \subseteq T(v_i)$ be a disc about $v_i^*$. Let $(\phi,\psi,g)$ be a CPN of $f$ along $c$. We have:

$$(*_{\text{open } c}) \qquad \begin{array}{ccc} I_i \times T(c) & \xrightarrow{(K_i\times\beta_i) \,=\, \phi_{v_i}^{-1}\circ\phi} & I \times T(v_i) \\ \Big\downarrow 1\times g & & \Big\downarrow 1\times g_{v_i} \\ I_i \times J & \xrightarrow{(K_i\times\gamma_i) \,=\, \psi_{v_i}^{-1}\circ e\circ\psi} & I \times J \end{array}$$

where $K_i = \pm L_i$.

Lemma (using the notation just introduced). Given a commutative diagram of maps:

$$\begin{array}{ccc} I_i \times \mathbb{D}' & \xrightarrow{K_i \times \beta_i'} & I \times \mathbb{D}_i^* \\ \Big\downarrow 1\times k_i & & \Big\downarrow 1\times g_{v_i} \\ I_i \times J' & \xrightarrow{K_i \times \gamma_i'} & I \times J \end{array}$$

where $K_i \times \beta_i'$ and $K_i \times \gamma_i'$ are embeddings, there is a CPN of $f$ along $c$, $(\phi',\psi',g')$ such that in a sufficiently small neighborhood $V_i$ of $I_i' \times \{0\} \subseteq I_i \times \{0\}$. $g' \mid V_i = k_i \mid V_i$, $\phi_{v_i}^{-1}\circ\phi' \mid V_i = K_i \times \beta_i' \mid V_i$ and on a sufficiently small neighborhood $W_i$ of $I_i' \times \{0\} \subseteq I_i \times \{0\}$, $\psi_{v_i}^{-1}\circ e\circ\psi' \mid W_i = K_i \times \gamma_i' \mid W_i$. (Here if $I_i = I_i(\alpha)$, then $I_i' = I_i(\alpha')$ for $0 < \alpha' < \alpha$.)

*Proof.* This proof is analogous to that of the preceding proposition, using Corollary 3 instead of Lemma 0 of §2.1. //

Applying this lemma we get:

*Proposition 2* (for $c$ open). *Let* $c \in C$ *be open, joining vertex* $v_-$ *to vertex* $v_+$. *There is a* CPN *of* $f$ *along* $c$, $(\phi,\psi,g)$ *so that for a sufficiently small neighborhood* $V_i$ *of* $I_i \times 0$ *in* $I_i \times T(c)$ *the maps in*

$$(*_{\text{open } c}) \qquad \begin{array}{ccc} I_i \times T(c) \supseteq V_i & \xrightarrow{\ \phi_{v_i}^{-1} \circ \phi \,=\, K_i \times \beta_i\ } & I \times \mathbb{D}_i^* \subseteq I \times T(v_i) \\ \Big\downarrow{\scriptstyle 1 \times g} & & \Big\downarrow{\scriptstyle 1 \times g_{v_i}} \\ I_i \times J \supseteq W_i & \xrightarrow{\ \psi_{v_i}^{-1} \circ e \circ \psi \,=\, K_i \times \gamma_i\ } & I \times J \end{array}$$

*are those given in the following catalogue:*

*Notation:* If $v$ is a vertex of type $(1\cdot2\cdot2\cdot1)$ or $(1\cdot2\cdot2\cdot3)$ we adopt the following notations for naming the arcs abutting $v$, and its singular $q$-preimages, $v^{\pm}$. $I$ in the CPN of $f$ at $v$ is oriented from left to right.

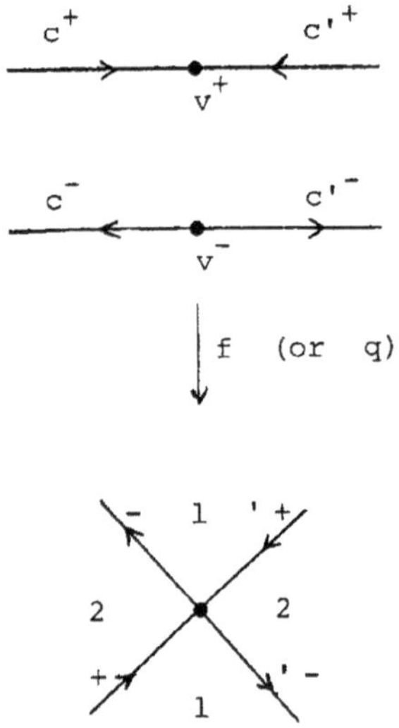


v of type (1·2·2·1)

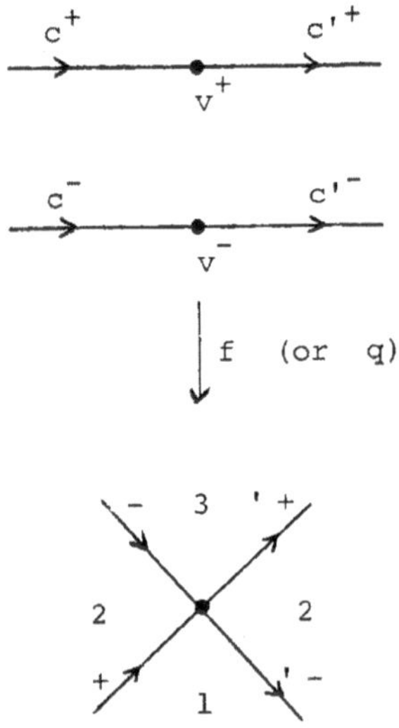


v of type (1·2·2·3)

The coordinates of $P^{\pm}$ are $(\pm x_0, 0)$ where $\phi_V(0,P^{\pm}) = v^{\pm}$. To make it clear which abutting arc we're working on, we label the CPN of $f$ along $c^{\pm}$ as $(\phi_{c^{\pm}}, \psi_{c^{\pm}}, g_{c^{\pm}})$ etc.; and picture the $f$-images of these arcs near $\bar{f}(v)$ on the left of each square of maps.

1) If $v$ is of type $(1 \cdot 2 \cdot 2 \cdot 1)$:

In a neighborhood of $\phi^{-1}_{c^{\pm}}(c^{\pm} \cap B(v^{\pm}))$:

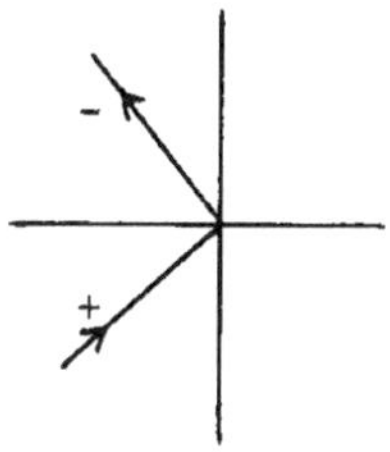

$$\begin{array}{ccc} u,(x,y) & \longrightarrow \pm L_{\pm}(u),(x \pm x_0,\ y) & \bar{u},(\bar{x},\bar{y}) \\ \downarrow & & \downarrow \\ u,(x^2-y^2) & & \bar{u},\pm\{(x \mp x_0)^2-\bar{y}^2+\bar{u}\} \\ (U,X) & \longrightarrow (\pm L_{\pm}(U), \pm(X \pm L_{\pm}(U))) & \end{array}$$

In a neighborhood of $\phi^{-1}_{c'^{\pm}}(c'^{\pm} \cap B(v^{\pm}))$:

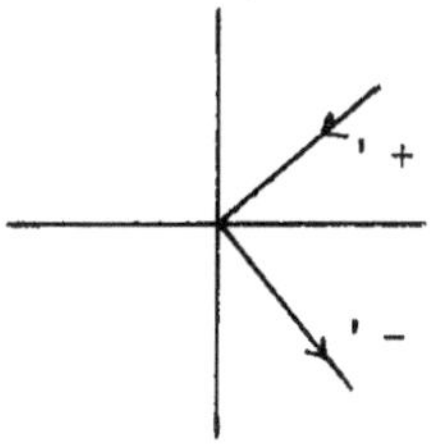

$$\begin{array}{ccc} u,(x,y) & \longrightarrow \pm(-L_{\pm}(u)),(y \pm x_0, \pm x) & \bar{u},(\bar{x},\bar{y}) \\ \downarrow & & \downarrow \\ u,x^2-y^2 & & u,\pm\{(\bar{x} \mp x_0)^2-\bar{y}^2+\bar{u}\} \\ (U,X) & \longrightarrow \pm(-L_{\pm}(U)),\pm\{-(X+L_{\pm}(U))\} & \end{array}$$

2) If $v$ is of type $(1 \cdot 2 \cdot 2 \cdot 3)$:

In a neighborhood of $\phi^{-1}_{c^{\pm}}(c^{\pm} \cap B(v^{\pm}))$:

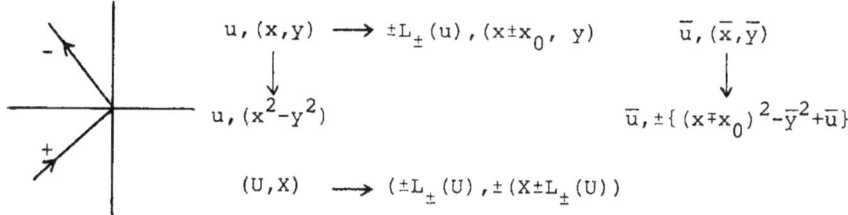

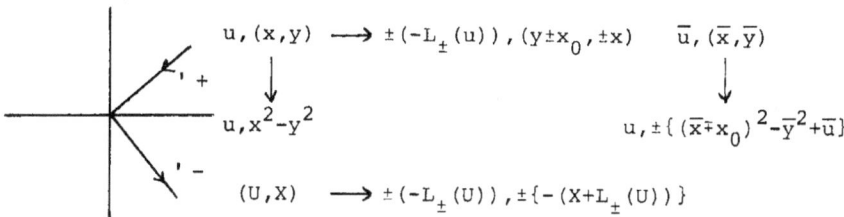

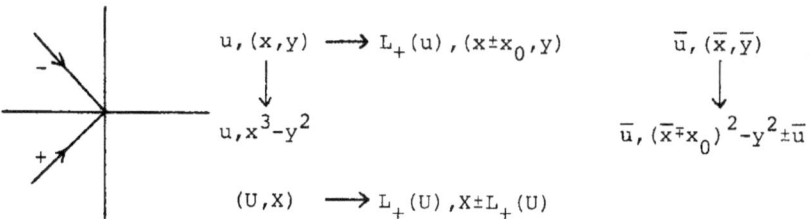

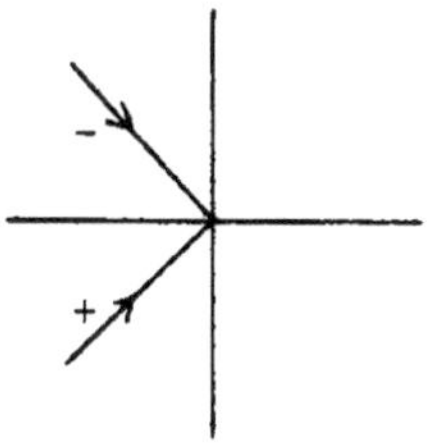

$$\begin{array}{ccc} u,(x,y) & \longrightarrow L_{+}(u),(x \pm x_0, y) & \bar{u},(\bar{x},\bar{y}) \\ \downarrow & & \downarrow \\ u,x^3-y^2 & & \bar{u},(\bar{x} \mp x_0)^2-y^2 \pm \bar{u} \\ (U,X) & \longrightarrow L_{+}(U), X \pm L_{+}(U) & \end{array}$$

In a neighborhood of $\phi^{-1}_{c'^{\pm}}(c'^{\pm} \cap B(v^{\pm}))$:

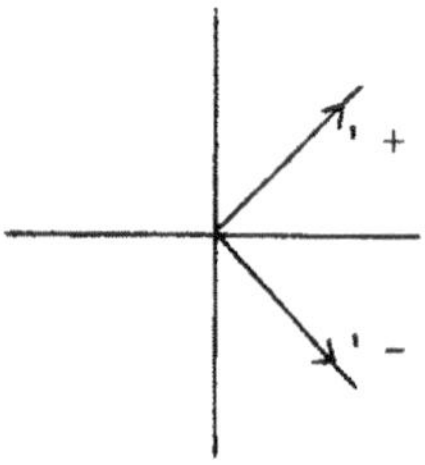


$$\begin{array}{ccc} u,(x,y) & \longrightarrow & L_-(u),(x\pm x_0,y) \qquad \bar{u},(\bar{x},\bar{y}) \\ \downarrow & & \downarrow \\ u,x^2-y^2 & & \bar{u},(\bar{x}\mp\bar{x}_0)^2-y^2\pm\bar{u} \\ (U,X) & \longrightarrow & L_-(U),X\pm L_-(U) \end{array}$$

3) If $v$ is a cusp, there are two cases to consider (see Prop. 1 of §2.2) namely those for which $g_v(u,(x,y)) = y^2 - xu + \varepsilon x^3$, $\varepsilon = \pm 1$. The singular set of $1 \times g_v$, $S(1\times g_v) = \{y = 0,\ u = 3\varepsilon x^2\}$ has the cusp at the origin and $S(1\times g_v) - \{\text{cusp}\}$ breaks up into two arcs $\{y = 0,\ x = \delta\sqrt{\varepsilon u/3}\}$ $\delta = \pm 1$. The arc is definite if $\varepsilon\delta = 1$ and indefinite if $\varepsilon\delta = -1$. The following graphs of the arcs abutting the cusp and their $g_v$ images give our coordinate and orientation conventions in the two cases $\varepsilon = \pm 1$.

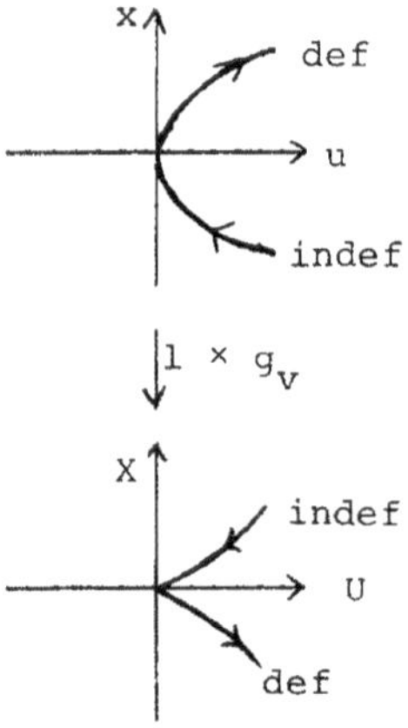


$\varepsilon = +1$

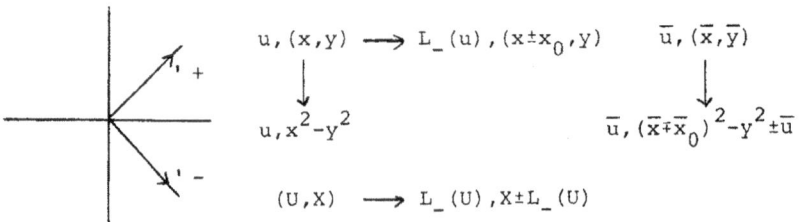

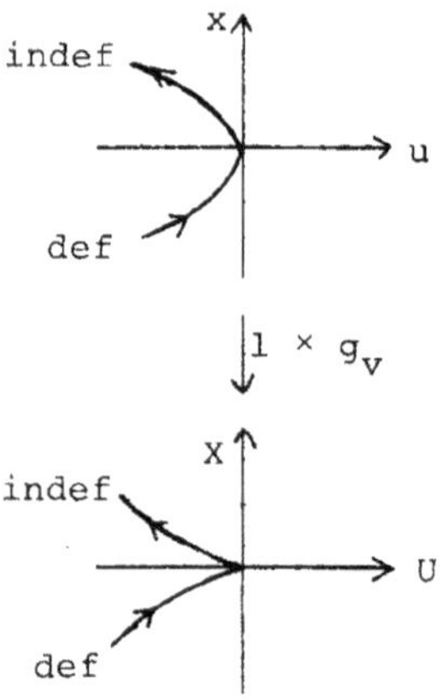


$\varepsilon = -1$

In a neighborhood of $\phi^{-1}_{c_{(\varepsilon,\delta)}}(c_{(\varepsilon,\delta)} \cap B(v))$:

$$\begin{array}{ccc} u,(x,y) & \longrightarrow & \varepsilon\delta L_{-\delta}(u),(\delta L_{-\delta}(u))^{1/2}(\tilde{x}+(\delta/\sqrt{3})),(\delta L_{-\delta}(u))^{3/4}\varepsilon\delta y \\ & & \bar{u},(\bar{x},\bar{y}) \\ \downarrow & & \downarrow \\ u,x^2+\varepsilon\delta y^2 & & \bar{u},\bar{y}^2 - \bar{x}\bar{u} + \bar{x}^3 \\ (U,X) & \longrightarrow & \varepsilon\delta L_{-\delta}(U),\varepsilon\delta(\delta L_{-\delta}(U))^{3/2}(X - (2/3\sqrt{3})) \end{array}$$

Here $x^2 = \tilde{x}^2(\sqrt{3} + \delta\tilde{x})$ defines $\tilde{x}$ uniquely if we require $x\tilde{x} \geq 0$. Notice that the horizontal maps are well defined and are diffeomorphisms since on their domains $\delta L_{-\delta} > 0$.

Note. From the forms of the maps given above if we replace the $\beta_i^*$ by $\beta_i^* \circ \rho$ where $\rho(u,(x,y)) = (u,(-x,-y))$, all the diagrams remain commutative. The proof applies equally well to these maps.

Without disturbing the canonical nature obtained from Proposition 2 for CPN of $f$ along $c$, $(\phi,\psi,g)$ near $(I_i \times 0)$, we show we can get a CPN of $f$ along $c$ such that $g$ has a canonical germ along all of $I \times 0$.

Proposition 3. *Let $c \in C_i$, $i = 0, 1$ and $(\phi,\psi,g)$ be any CPN of $f$ along $c$. Then there is a $(\phi',\psi',g')$, (CPN) of $f$ along $c$ such that $\psi' = \psi$ and such that in a sufficiently small neighborhood $U$ of $I \times \{0\}$ in $I \times T(c)$,*

$$g' \mid U = x^2 + (-1)^i y^2.$$

*Further if on any set $W \subseteq I \times T(c)$, $g \mid W = x^2 + (-1)^i y^2$ then there is a neighborhood $W' \subseteq W$ of $W \cap (I\times 0)$ such that on $W'$, $g' = g$ and $\phi' = \phi$ or $\phi' = \phi\circ\rho$, where $\rho(u,\xi) = (u,-\xi)$.*

(Note. The possibility that $\phi' = \phi\circ\rho$ can only arise if $c \in C_1$.)

Proof. By Corollary 1 of §2.1 it suffices to show:

Lemma. *Let $g : I \times \mathbb{D} \longrightarrow \mathbb{R}$ be a smooth map such that $g \mid I \times 0 = 0$, $(\partial g/\partial x) \mid I \times 0 = (\partial g/\partial y) \mid I \times 0 = 0$ and the Hessian of $g$ with respect to $x$ and $y$ is non-singular everywhere on $I \times 0$. Then there is an orientation preserving diffeomorphism $1 \times \phi$ of a neighborhood of $I \times 0$ in $I \times \mathbb{D}$ such that in that neighborhood*

$$g\circ(1\times\phi)(u,(x,y)) = x^2 - y^2 \quad (\text{or } \pm(x^2 + y^2))$$

*if the Hessian of $g$ is indefinite (or definite). Further if at any $(u,0) \in I \times 0$, the germ of $g$ has one of the forms $(x^2 + y^2)$ (or $(x^2 - y^2)$), then the germ of $(1\times\phi)$ at $(u,0)$ is the germ of the identity map (or the identity map or $\rho$).*

Remark. In the lemma there are two canonical forms for the definite case whereas in the proposition there is only one. This is because our choice of orientation requires that in the definite case, $g(u,(x,y))$ have a minimum at $(u,0)$, so the Hessian is positive definite.

Proof of the lemma.
Write $g(u,(x,y)) = ax^2 + 2bxy + cy^2 + kux + \ell uy$ where $k$ and $\ell$ are functions of $u$ only. Since $g_x(u,0) = g_y(u,0) = 0$, both $k$ and $\ell$ vanish. Thus it is no restriction to write $g(u,\xi) = A(u,\xi)\xi^t$ where $\xi = (x,y)$. The Hessian of $g$ with respect to $(x,y)$ at $(u,0)$ is $2A(u,0)$. There are two cases:

(i) $A(u,0)$ is indefinite. In a neighborhood $U$ of $I \times 0$ in $I \times \mathbb{D}$, $A$ is indefinite, hence the two eigenvalues define two smooth functions $\lambda_i : U \longrightarrow \mathbb{R}$ such that $\lambda_1\lambda_2 < 0$ on $U$. Suppose $\lambda_1 > 0$. Let $p : U \longrightarrow \mathbb{R}^2$ be a $\lambda_1$-eigenvector of $A$ at each point of $U$, with $|p| = 1$. Let $p^\perp : U \longrightarrow \mathbb{R}^2$ be such that $|p^\perp| = 1$ and $p\cdot p^\perp = 0$ and $p \wedge p^\perp = \det P = 1$ where $P$ is the orthogonal matrix with rows $p$ and $p^\perp$. Obviously $p^\perp$ is a $\lambda_2$-eigenvector of $A$ at each point of $U$. We write

$$g(u,\xi) = \xi P^t(u,\xi)P(u,\xi)A(u,\xi)P^t(u,\xi)P(u,\xi)\xi^t$$

$$= \xi P^t(u,\xi)\begin{pmatrix}\lambda_1(u,\xi) & 0\\ 0 & \lambda_2(u,\xi)\end{pmatrix}P(u,\xi)\xi^t.$$

Let $\Delta = \begin{pmatrix}\lambda_1 & 0\\ 0 & \lambda_2\end{pmatrix}$. Define the map

$$1 \times \theta : U \longrightarrow I \times \mathbb{D} : (u,\xi) \longrightarrow (u,\xi P^t(u,\xi)).$$

Since the determinant of the Jacobian of $1 \times \theta$ is positive on $I \times 0$, we know we can cover $I \times 0$ with little balls on each of which $1 \times \theta$ is a diffeomorphism. On the union of these balls $(1\times\theta)$ is obviously 1:1. Thus, on a neighborhood of $I \times 0$, $1 \times \theta$ is an orientation preserving diffeomorphism. Let $1 \times \psi = (1\times\theta)^{-1}$, and let $\lambda_1\circ(1\times\psi) = \alpha^2$ and $\lambda_2\circ(1\times\psi) = -\beta^2$ where $\alpha > 0$ and $\beta > 0$. Since $g = \theta\cdot\Delta\cdot\theta^t$, $g\circ(1\times\psi)(u,(x,y)) = x^2\alpha^2(u,x,y) - y^2\beta^2(u,x,y)$. Letting $\vartheta(u,x,y) = (x\alpha(u,x,y), y\beta(u,x,y))$, the diffeomorphism $1 \times \phi = (1\times\psi)\circ(1\times\vartheta)^{-1}$ is required diffeomorphism.

If in a neighborhood of $(u_0,0)$, $g(u,(x,y)) = x^2 - y^2$, $P = \pm I$ in that neighborhood, hence $\phi(u,\xi) = \pm\xi$.

(ii) $A(u,0)$ is positive definite. (The modification for the negative definite case is obvious.) Let $U$ be a neighborhood of

$I \times 0$ on which $A$ is positive definite. Choose $q : U \longrightarrow \mathbb{R}^2$ so that $|q| = 1$, and $qA \perp (1,0)$ on $U$. Since $A$ is positive definite $q$ cannot be a multiple of $(1,0)$, so we can require that $(1,0) \wedge q > 0$. Letting $P$ be the matrix whose rows are $(1,0)$ and $q$ we have a mapping, $P : U \longrightarrow SL(2,\mathbb{R})$ such that

$$PAP^t = \Delta = \begin{pmatrix} \alpha^2 & 0 \\ 0 & \beta^2 \end{pmatrix}$$

where $\beta^2 = qAq^t$ and $\alpha^2 = a_{11}$ (11-entry of $A$), where $\alpha > 0$ and $\beta > 0$. As in the previous case, let $\theta(u,(x,y)) = (x,y)P^{-1}(u,(x,y))$. Letting $1 \times \psi = (1\times\theta)^{-1}$, we have:

$$g\circ(1\times\psi)(u,(x,y)) = x^2\alpha^2(u,\psi(u,x,y)) + u^2\beta^2(u,\psi(u,x,y)).$$

Let $\vartheta(u,(x,y)) = x\alpha(u,\psi(u,x,y)), \ y\beta(u,\psi(u,x,y))$. The required $1 \times \phi = 1 \times \psi\circ(1\times\vartheta)^{-1}$.

If in a neighborhood of $(u_0,0)$, $g(u,(x,y)) = x^2 + y^2$, then $P = I$ and hence $\phi(u,(x,y)) = (x,y)$ in that neighborhood. //

*Theorem.* *Let* $c \in C_j$, $j = 0, 1$ *be either closed or open joining vertex* $v_-$ *to* $v_+$. *Then there is a CPN of* $f$ *along* $c$, $(\phi,\psi,g)$ *such that on a sufficiently small neighborhood* $U$ *of* $I \times 0 \subseteq I \times T(c)$, $g \mid U\ (u,(x,y)) = x^2 + (-1)^j y^2$. *Further on neighborhoods of* $I_+ \times 0 \subseteq I \times T(c)$ *for* $c$*-closed or on neighborhoods of* $I_i \times 0 \subseteq T(c)$, $i = +,-$ *for* $c$*-open, the maps in the commutative squares* $(*_{\text{closed } c})$ *and* $(*_{\text{open } c})$ *are those given by both parts of Proposition 2.*

We treat both cases simultaneously calling the neighborhoods of $I_i \times 0$, $V_i$.

*Proof.* All that is needed here is to deal with the case of $c \in C_1$ such that when the map $\phi$ constructed in Proposition 2 is replaced by $\phi'$ in the construction of Proposition 3, $\phi' = \phi$ or $\phi' = \phi\circ\rho$ on $V_i$. Obviously if $\phi' = \phi\circ\rho$ at both ends, we merely change the choice of $p$ in part (i) of the construction of Proposition 3 to $-p$. Thus we suppose $\phi' = \phi$ on $V_-$ and $\phi' = \phi\circ\rho$ on $V_+$.

Let $\vartheta : I \longrightarrow [0,\pi]$ be non-decreasing with $\vartheta \mid I_- = 0$ and $\vartheta \mid I_+ = \pi$. Let $\tau : I \times \mathbb{R}^2 \longrightarrow \mathbb{R}^2 : (u,z) \longrightarrow e^{i\vartheta(u)}z$. (Here we identify $\mathbb{R}^2$ with $\mathbb{C}$.) $(1\times\tau) \mid I_- \times \mathbb{R}^2$ is the identity map and $(1\times\tau) \mid I_+ \times \mathbb{R}$ is $\rho$. For any nested discs about $0$, $\mathbb{D}' \subseteq \mathbb{D} \subseteq T(c)$, use Lemma 0 of §2.1 to extend $(1\times\tau) \mid I \times \mathbb{D}'$ to a diffeomorphism $1 \times \tau'$ of $I \times T(c)$. Define a new CPN of $f$ along $c$ as $\phi' \circ (1\times\tau')$, $\psi'$, $g' \circ (1\times\tau')$. Restricted to neighborhoods $V_i'$, this CPN gives the required mappings. We now apply the construction of Proposition 3 to this CPN. Along $I \times 0$, the germ of $g' \circ (1\times\tau)$ is just

$$(x \cos \vartheta - y \sin \vartheta)^2 - (x \sin \vartheta + y \sin \vartheta)^2 = (x,y)A(u)(x,y)^t,$$

where $A(u) = \begin{pmatrix} \cos 2\vartheta & -\sin 2\vartheta \\ -\sin 2\vartheta & -\cos 2\vartheta \end{pmatrix}$. If, in the notation of the proof of Proposition 3, we choose as $p$, the eigenvector belonging to $1$ of $A(u)$ as: $(\cos \vartheta, -\sin \vartheta)$, we see that $p \mid I_i \times 0 = (1,0)$. Hence the germ of $P$ on $I_i \times 0$ is the identity matrix and thus the germ of the coordinate change that produces the new CPN is the identity along $I_i \times 0$. //

Note. From now on the CPN of $f$ at $v$ for all $v \in V$ and the CPN of $f$ along $c$ for all $c \in C$ are fixed. They are chosen from among those described in Proposition 1 of §2.2 and the preceding theorem. They will be denoted by $(\phi_v, \psi_v, g_v)$ and $(\phi_c, \psi_c, g_c)$ and will be referred to as canonical coordinatized product neighborhoods or $C^2$PN.

CHAPTER III

# Lifting Stable Maps of Three Manifolds into the Plane to Immersions in $\mathbb{R}^4$



## 3.0 Introduction and summary of the results

In this chapter we investigate the following question:

I. *Given a stable map $f : M \longrightarrow \mathbb{R}^2$, $M$, a compact, oriented, three dimensional manifold, does there exist an $h : M \longrightarrow \mathbb{R}^2$ such that $(f,h) : M \longrightarrow \mathbb{R}^4$ is an immersion.*

Denote by $I_f(M,\mathbb{R}^4)$ the set of all immersions of $M$ in $\mathbb{R}^4$ over $f$. We say that two such $(f,h_0)$ and $(f,h_1)$ are f-*regularly homotopic* if there is a homotopy $H : M \times I \longrightarrow \mathbb{R}^2$ such that $(f,h_t) \in I_f(M,\mathbb{R}^4)$ for all $t \in I$. (Here $H(x,t) = h_t(x)$.) We denote by $[I_f(M,\mathbb{R}^4)]$ the set of f-regular homotopy classes of immersions of $M$ in $\mathbb{R}^4$ over $f$. More general than the first question is:

$I_1$. <u>Given a stable map</u> $f : M \longrightarrow \mathbb{R}^2$, $M$, <u>a compact oriented, three dimensional manifold, describe (classify) the set</u> $[I_f(M,\mathbb{R}^4)]$.

It will sometimes be convenient to consider the sets $F_f(M,\mathbb{R}^2) = \{h : M \longrightarrow \mathbb{R}^2 \mid (f,h) \in I_f(M,\mathbb{R}^4)\}$. We call $F_f(M,\mathbb{R}^2)$ the set of <u>f-fibre immersions of</u> $M$ in $\mathbb{R}^2$. We say that $h_0$ and $h_1$ in $F_f(M,\mathbb{R}^2)$ are <u>f-fibre regularly homotopic if</u>, $[f,h_0] = [f,h_1] \in [I_f(M,\mathbb{R}^4)]$, and we denote the set of such homotopy classes by $[F_f(M,\mathbb{R}^2)]$.

An element $h \in F_f(M,\mathbb{R}^2)$ can also be described as a smooth map from $M$ to $\mathbb{R}^2$ such that for each $x \subseteq M$, if $K_x = \ker Tf_x \subseteq TM_x$, then $Th_x \mid K_x$ is injective.

Trivially $I_f(M,\mathbb{R}^4)$ and $F_f(M,\mathbb{R}^2)$ (and $[I_f(M,\mathbb{R}^4)]$ and $[F_f(M,\mathbb{R}^2)]$) are in 1:1 correspondence. Thus any notion and notation introduced for one has an obvious interpretation and analog for the other.

Since $f$ is fixed, we will suppress the subscript $f$ throughout. We will also abbreviate $I_f(M,\mathbb{R}^4)$ and $F_f(M,\mathbb{R}^2)$ to $I$ and $F$.

Let $B(\mathcal{R}) = \bigcup_{R\in\mathcal{R}} B(R)$; this is a trivial circle bundle over $\cup\{R \mid R \in \mathcal{R}\}$. By a result of Eliasberg and Gromov [E-G], $[F(B(\mathcal{R}),\mathbb{R}^2)]$ is in 1:1 correspondence with the homotopy classes of maps of $B(\mathcal{R})$ into $\mathbb{R}^2 - \{0\}$ where the class of $h$, $[h]$, is taken to the homotopy class of the map that takes a point $x \in B(\mathcal{R})$ to the $Th_x$-image of the orientation vector of the fibre of $B(\mathcal{R})$ through $x$. The homotopy classes of maps of $B(\mathcal{R})$ into $\mathbb{R}^2 - \{0\}$ are in turn in 1:1 correspondence with $H^1(B(\mathcal{R}))$ via the induced map on homology. Thus we have a map

$$A : [I] \longrightarrow H^1(B(\mathcal{R})) \quad \text{as the composition}$$

$$[I] \longrightarrow [I(B(\mathcal{R}),\mathbb{R}^4)] \longleftrightarrow H^1(B(\mathcal{R}))$$

where the first map is projection by restricting $(f,h)$ to $B(\mathcal{R})$. If we let $r_h(R)$ be the value of $A[f,h]$ on the orientation class of the fibre of $B(R)$, we characterize those elements of $\mathbb{Z}^{\mathcal{R}}$ which arise as $r_h$ for some $[h] \in F$, (see §3.2); they are those elements of $\mathbb{Z}^{\mathcal{R}}$ which determine an orientation of $S(f)$. We call the set of such functions, $\text{Rot} \subseteq \mathbb{Z}^{\mathcal{R}}$.

Choosing trivializations $\phi_R : R \times S^1 \longrightarrow B(R)$ for each $R \in \mathcal{R}$, we can write $\phi_R^* \circ A = (r,s)$:

$$[\mathcal{I}] \longrightarrow (\underset{R\in\mathcal{R}}{\times} \mathbb{Z}) \times (\underset{R\in\mathcal{R}}{\times} H^1(R))$$

$$[f,h] \longrightarrow ( \quad r_h \quad , \quad s_h \quad ).$$

Since the trivializations are chosen once and for all and fixed we will suppress the $\phi_R^*$. By choosing generators we may write $H^1(R)$ as $\mathbb{Z}^{2m(R)} \times \mathbb{Z}^{n(r)-1}$, where $m(R)$ is the number of handles of $R$ and $n(R)$ is the number of boundary components of $R$.

Thus we can write a more explicit version of $A$:

$$A : [\mathcal{I}] \longrightarrow \mathrm{Rot} \times (\underset{R\in\mathcal{R}}{\times} \mathbb{Z}^{2m(R)}) \times (\underset{R\in\mathcal{R}}{\times} \mathbb{Z}^{n(R)-1})$$

$$[f,h] \longrightarrow ( r_h , \quad a_h \quad , \quad b_h \quad )$$

In the next sections we focus on the f-regular homotopy classes of the restrictions of $(f,h)$ to a neighborhood $B(\Sigma)$ of $\hat{\Sigma} = q^{-1}(\Sigma)$. A first invariant of elements of $[\mathcal{I}]$ is a map $\delta : \mathcal{C} \longrightarrow \{+1,-1\}$. If $(f,h) \in \mathcal{I}$, and $c \in \mathcal{C}$, $\delta_h(c) = +1$ (or $-1$) if Th restricted to the tangent space to the transverse manifold at any point $x \in q^{-1}(c) \cap S(f)$ is orientation preserving (or reversing). We denote by $\mathcal{D}$ the set of maps $(\mathcal{C},\{+1,-1\})$ satisfying obvious consistency conditions (given in §3.4); it is clear that any $\delta \in \mathcal{D}$ can arise from a <u>germ at</u> $S(f)$ of an immersion $(f,h)$. It is not clear that every $\delta$ is $\delta_h$ for some immersion $(f,h)$ of all of $M$ in $\mathbb{R}^4$.

In order to complete the classification of the f-regular homotopy classes of germs of maps $\hat{\Sigma}$ which arise as restrictions of elements of $\mathcal{I}$ we "standardize" the elements of $\mathcal{I}$ in several stages. We first show in §3.3 that if we choose any embedding or immersion, $E$, of $S(f)$ in $\mathbb{R}^4$ over $f \mid S(f)$, that $[\mathcal{I}] = [\mathcal{I}_E]$, where $\mathcal{I}_E = \{(f,h) \in \mathcal{I} \mid (f,h) \mid S(f) = E\}$. Furthermore if $\mathcal{I}_V = \{(f,h) \in \mathcal{I}_E \mid$ the germ of $h$ at $v$ for all $v \in V$ is defined in terms of $\delta_h\}$, then in §3.4 we show that $[\mathcal{I}] = [\mathcal{I}_V]$ as well. This standardization of the germ of $h$ at $V$, $(h)_V$ is effected using the $C^2PN$ of §2.2.

In §3.6 we define the next invariant for which we first construct a one-dimensional subcomplex $\Sigma^* \subseteq \hat{\Sigma}$. On each closed component of $S(f)$ choose a point, $v^c$. Let $V^* \subseteq S(f)$ consist of the set of $v^c$ points and the set $q^{-1}(V) \cap S(f)$ for $V$ the set of vertices of $f$, and let $V_* = q(V^*)$. Let $\Sigma^* = q^{-1}(V_*) \cup S(f)$. We consider $\Sigma^*$ as a one-dimensional cell complex with vertices, $V^*$ and open 1-cells $q^{-1}(c) \cap S(f) - V^*$ and $q^{-1}(v) - S(f)$ for all $c \in \mathcal{C}$ and $v \in V_*$.

All of these 1-cells are oriented (we use the orientation of M to define these orientations). For any $(f,h) \in I_{V_*}$ (where the germ of h at $v \in V^*$ is defined in terms of $\delta_h$} we define an element of $C^1(\Sigma^*,\mathbb{R})$. Since h immerses all of the arcs of $q^{-1}(v) - S(f)$ for any $v \in V_*$, if $\alpha$ is such an arc, we let $2\pi\sigma_h(\alpha)$ be the total turning of the h image of the positively oriented tangent to $\alpha$. On any arc $\hat{c} = q^{-1}(c) \cap S(f) - V^*$, we let $2\pi\tau_h(c)$ be the total turning of the h-image of the tangent space to the transverse manifold to $\hat{c}$ as we travel in the positive sense from the initial vertex of $\hat{c}$ to its final vertex. Amalgamating these two functions we obtain a map $\tau_h \cup \sigma_h \in C^1(\Sigma^*,\mathbb{R})$. We define an integer valued $\nu_h$ such that $\nu_h - (\tau_h \cup \sigma_h) \in [-1/2,1/2]$. In §3.11 it is shown that the cohomology class of $\nu_h$, $[\nu_h] \in H^1(\Sigma^*)$ depends only on the class of (f,h) in $[I]$. Thus we have a map $B : [I] \longrightarrow (\mathcal{D} \times H^1(\Sigma^*)) : [f,h] \longrightarrow (\delta_h,[\nu_h])$, whose value determines f-regular homotopy class of the germ of (f,h) at $\hat{\Sigma}$ for any $[f,h] \in [I]$ (see §3.6 Proposition 1). In §3.7 we describe the subset $N' \subseteq C^1(\Sigma^*,\mathbb{Z})$ for any $\delta \in \mathcal{D}$ such that if $\nu \in N'_\delta$ then there is an immersion germ $(f,h)_{\hat{\Sigma}}$ with $(\delta,\nu) = (\delta_h,\nu_h)$. In §3.8 we describe the additional restrictions that produce a smaller set $N^*_\delta \subseteq N'_\delta$ so that for any $\nu \in N^*_\delta$ there is an $(f,h) \in I_{V_*}$ with $(\delta,\nu) = (\delta_h,\nu_h)$; let $N_\delta \subseteq H^1(\Sigma^*)$ be the image of $N^*_\delta$. We also show there that two of the components of the map A, $r_h$ and $b_h$ are determined by B. To this point the value $(A,B)[f,h]$ determines the f-regular homotopy class of $(f,h) \mid B(R)$ and of the germ $(f,h)_{\hat{\Sigma}}$. To complete the classification we must show how to distinguish elements in the fibre of (A,B). To this end, for each p in the image of (A,B) choose a representative $[f,h_p] \in [I]$ of $(A,B)^{-1}(p)$. For any other $[f,h] \in (A,B)^{-1}(p)$ we compare $[f,h]$ with our standard $[f,h_p]$ by choosing representatives h and $h_p$ that agree on a neighborhood $B(\Sigma)$ of $\hat{\Sigma}$ and forming the obvious map $h \cdot h_p$ on the double of $M - B(\Sigma)$, $(M - B(\Sigma))^{(2)}$ into $\mathbb{R}^2$. This double is essentially the union of the doubles, $B(R)^{(2)}$ for all $R \in \mathcal{R}$, so it is a trivial circle bundle over the union of the doubles, $R^{(2)}$ for all $R \in \mathcal{R}$. Since $h \cdot h_p$ immerses each circle fibre in $\mathbb{R}^2$, we construct a map (in §3.9 and 3.11), $C : [I] \longrightarrow \underset{R \in \mathcal{R}}{\times} H^1(R^{(2)})$ analogous to the $s_h$ component of the map A. The map analogous to the $r_h$-component of A gives nothing new and in fact most of the information in $C[f,h]$ comes from $s_h$ as well. However there is enough that is new to give:

Theorem. $(A,B,C) : [I] \longrightarrow \underset{R\in\mathcal{R}}{\times} (\mathbb{Z} \times H^1(R)) \times (\mathcal{D} \times H^1(\Sigma^*)) \times (\underset{R\in\mathcal{R}}{\times} H^1(R^{(2)}))$ is injective.

If, as we did before, we choose generators in $H^1(R)$ and $H^1(R^{(2)})$, we write $H^1(R) = \mathbb{Z}^{2m(R)} \times \mathbb{Z}^{n(R)-1}$ and $H^1(R^{(2)}) = \mathbb{Z}^{2m(R)} \times \mathbb{Z}^{2m(R)}$ $\mathbb{Z}^{n(R)-1} \times \mathbb{Z}^{n(R)-1}$ we have

$$(A,B,C) : [I] \longrightarrow \mathrm{Rot} \times \underset{R\in\mathcal{R}}{\times} (\mathbb{Z}^{2m(R)} \times \mathbb{Z}^{n(R)-1}) \times \mathcal{D} \times H^1(\Sigma^*) \times$$

$$\underset{R\in\mathcal{R}}{\times} (\mathbb{Z}^{2m(R)} \times \mathbb{Z}^{2m(R)} \times \mathbb{Z}^{n(R)-1} \times \mathbb{Z}^{n(R)-1})$$

$$[f,h] \longrightarrow (r_h, a_h, b_h, \delta, [\nu_h], a_h, a_h, b_h, c_h)$$

where the $c_h$ factors comes from the subspace of $H_1(R^{(2)})$ generated by the doubles of arcs joining components of the boundary of $R$. Since $r_h$ and $b_h$ are determined by $\delta_h$ and $\nu_h$ we have a more compact statement. If we let $D$ be the projection of $(A,B,C)$ on the appropriate components we have:

Theorem. $D : [I] \longrightarrow (\bigcup_{\delta\in\mathcal{D}} (\delta \times N_\delta) \times (\underset{R\in\mathcal{R}}{\times} H^1(R,\partial R))$ is a 1:1 correspondence where $D[f,h] = (\delta_h, [\nu_h]) \times (a_h, c_h)$.

In the course of the following sections we consider more and more restricted kinds of immersions and equivalences. We have already introduced $I_E$ as the subset of $I$ such that $(f,h) \in I_E$ iff $(f,h) \mid S(f) = E$, a fixed embedding or immersion. If in addition we fix the germs of $h$ on $q^{-1}(V) \cap S(f)$, on $S(f)$ and on $\hat{\Sigma}$, we obtain $I_V \subseteq I_S \subseteq I_\Sigma$. The set of f-regular homotopy classes that respect those restrictions during the homotopies are denoted by $[I_E]_E$, $[I_V]_V$, $[I_S]_S$ and $[I_\Sigma]_\Sigma$ respectively. In §3.4, 3.5 and 3.6, we show that $[I_V]_V \longrightarrow [I_E]_E \longrightarrow [I]$ are both surjective and that $[I_\Sigma]_\Sigma \longrightarrow [I_S]_S \longrightarrow [I_V]_V$ are both 1:1 correspondences. We prove the classification theorem by first finding invariants of $[I_\Sigma]_\Sigma$ and then introducing into the space of invariants the equivalence relation induced by projecting $[I_\Sigma]_\Sigma$ into $[I]$.

Note. In this summary, since I was free of any obligation to prove the statements made, I presented the material in an order that seemed better for such a resume than the order of the text. I have also used a notation for the maps A, B, C, D which is free from any allusion to their histories; this also does not appear in the text. I hope that neither of these inconsistencies will cause the reader any trouble.

## 3.1 Necessary conditions on $h \mid B(R)$; the function $r_h$

Given an immersion $(f,h) : M \longrightarrow \mathbb{R}^4$, we derive some conditions on $W$ which are imposed by $h$. We work with $B(R)$, the circle bundle over $R$ for each $R \in \mathcal{R}$. We restrict ourselves to the oriented case. $B(R)$ is a trivial bundle for which we choose a trivialization $\phi_R : R \times S^1 \longrightarrow B(R)$, once and for all. Since $\bar{f} : R \longrightarrow \mathbb{R}^2$ is an immersion we orient $R$ so that $\bar{f}$ is orientation preserving and since $B(R)$ is oriented we orient the fibre, $S^1$, so that the orientation of $R$ followed by that of $S^1$ orients $B(R)$. The immersion $(f,h)$, immerses each fibre of $B(R)$ in $\mathbb{R}^2 \times \mathbb{R}^2$, but since $f$ takes a fibre of $B(R)$ to a point, we see that $h$ immerses each fibre of $B(R)$ into $\mathbb{R}^2$. Now since $R$ is connected the rotation number or the tangent degree of $h \mid B(R)_x$ is independent of $x \in R$. We denote this integer by $r_h(R)$. Thus $(f,h) \in \mathcal{I}$, determines a map $r_h : \mathcal{R} \longrightarrow \mathbb{Z}$.

Note. In the non-oriented case, the fibre circle of $B(R)$ has no natural orientation so that the analogous function could only be defined up to sign. However suppose $B(R)$ were not oriented, then there must be an embedded circle in $R$ over which sits a Klein bottle in $B(R)$. However if $(f,h)$ immersed $B(R)$ in $\mathbb{R}^4$, then the rotation number of $h$ on a circle fibre of $B(R)$ must vanish. This is eash to see for suppose $K$, a Klein bottle, is the identification space of $I \times S^1$ by identifying $0 \times S^1$ with $1 \times S^1$ with opposite orientations. Let $i : I \times S^1 \longrightarrow K$ be the identification map. Suppose $h : K \longrightarrow \mathbb{R}^2$ is an immersion on $i(t \times S^1)$ for all $t \in I$. Then for the rotation numbers we have $\mathrm{rot}(h \circ i \mid 0 \times S^1) = -\mathrm{rot}(h \circ i \mid 1 \times S^1)$. But since $\mathrm{rot}(h \circ i \mid t \times S^1)$ is independent of $t$, it must vanish. //

It is obvious that the map $r_h : \mathcal{R} \longrightarrow \mathbb{Z}$ is independent of the trivializations of $B(R)$ for $R \in \mathcal{R}$. Also if $[f,h] = [f,h'] \in [\mathcal{I}]$, then $r_h = r_{h'}$. Thus $r$ defines a map from $[\mathcal{I}]$ to $(\mathcal{R},\mathbb{Z})$, the maps from $\mathcal{R}$ to $\mathbb{Z}$. In the next paragraph we will develop necessary conditions for such a map to arise as $r_h$ for $h \in \mathcal{F}$.

Let $VB(R)$ be the kernel of $T(f \mid B(R))$, a line subbundle of $TB(R)$, the tangent bundle of $B(R)$. As already remarked $\mathcal{F}$ is the set of $h : M \longrightarrow \mathbb{R}^2$ whose tangent map, $Th$ has maximal rank on the kernel of $Tf$. Thus $\mathcal{F}(B(R),\mathbb{R}^2)$ is the set of all maps $h : B(R) \longrightarrow \mathbb{R}^2$ whose tangent map restricted to $VB(R)$ has maximal rank. By [E-G, Theorem 1.2.5], there is a 1:1 correspondence between the set of f-fibre regular homotopy classes, $[\mathcal{F}(B(R),\mathbb{R}^2]$, and the homotopy classes of maps of maximal rank, $[M(VB(R),T\mathbb{R}^2)]$.

Let $v : B(R) \longrightarrow VB(R)$ take $x \in B(R)$ to the unit vector orienting the fibre of $B(R)$ at $x$. Given any map $\mathfrak{h} = (\alpha,\beta) : VB(R) \longrightarrow \mathbb{R}^2 \times \mathbb{R}^2 = T\mathbb{R}^2$ of maximal rank, let $P_{\mathfrak{h}} : B(R) \longrightarrow S^1 : x \longrightarrow \beta(v(x))/|\beta(v(x))|$. Thus $[M(VB(R),T\mathbb{R}^2)]$ is in 1:1 correspondence with the set of homotopy classes $[B(R),S^1]$ of maps of $B(R)$ into $S^1$, which in turn is in 1:1 correspondence with the $H^1(B(R),\mathbb{Z})$.

Proposition. *$[I(B(R),\mathbb{R}^4)]$ and $H^1(B(R),\mathbb{Z})$ are in 1:1 correspondence via the map that takes $[(f,h)]$ to the homomorphism $(P_{Th})_* : H_1(B(R)) \longrightarrow H_1(S^1) = \mathbb{Z}$.*

Let $\phi_R : R \times S^1 \longrightarrow B(R)$ be our trivialization, then $(P_{Th}\circ\phi_R)_* : H_1(R) \oplus H_1(S^1) \longrightarrow \mathbb{Z}$ is the sum of two homomorphisms $s_h(R) : H_1(R) \longrightarrow \mathbb{Z}$ and $r_h(R) : H_1(S^1) \longrightarrow \mathbb{Z}$. Here we identify $r_h(R)$ with the image of the orientation class of $S^1$.

Corollary. *$[I(B(R),\mathbb{R}^4)]$ and $H^1(R,\mathbb{Z}) \times \mathbb{Z}$ are in 1:1 correspondence.*

Remark. *If $r_h(R) = 0$, then $s_h(R)$ is independent of the choice of trivialization, $\phi_R$.*

Proof. Given a trivialization $\phi_R$, we describe $s_h(R)$ explicitly. Let $\sigma : R \longrightarrow R \times S^1 : x \longrightarrow (x,1)$ then $s_h(R) = (P_{Th}\circ\phi_R\circ\sigma)_*$. (Here we consider $S^1 \subseteq \mathbb{C}$ as $\{z \in \mathbb{C} \mid |z| = 1\}$.) Thus if $\phi'_R$ is another trivialization we must compare $(P_{T_h}\circ\phi_R\circ\sigma)$ and $(P_{T_h}\circ\phi'_R\circ\sigma)$. Let $\psi = \phi_R^{-1}\circ\phi'_R : R \times S^1 \circlearrowleft : (x,z) \longrightarrow (x,\lambda(x,z))$, $\psi\circ\sigma : R \longrightarrow R \times S^1 : x \longrightarrow (x,\lambda\circ\sigma(x))$. Thus

$$(\psi\circ\phi)_* : H_1(R) \longrightarrow H_1(R) \oplus H_1(S^1) : a \longrightarrow \sigma_*(a) + (0,(\lambda\circ\sigma)_*(a))$$

hence

$$(\phi_R\circ\psi\circ\sigma)_*(a) = (\phi_R\circ\sigma)_*(a) + (\phi_R)_*(0,(\lambda\circ\sigma)_*(a)).$$

But $P_{Th_*}\phi_{R_*}(0,(\lambda\circ\sigma)_*(a)) = r_h(R)(\lambda\circ\sigma)_*(a)$. Thus if $r_h(R) = 0$, $(P_{T_h}\circ\phi_R\circ\psi\circ\sigma)_* = (P_{T_h}\circ\phi_T\circ\sigma)_*$. //

Gathering the information from the Proposition and Corollary together we have

Corollary. *$[I_f(\bigcup_{R\in\mathcal{R}} B(R),\mathbb{R}^4)]$ and $\times_{R\in\mathcal{R}} H^1(B(R)) \cong \times_{R\in\mathcal{R}} (H^1(R) \times \mathbb{Z})$ are in 1:1 correspondence. This correspondence is given by $[f,h] \longrightarrow (s_h,r_h)$.*

## 3.2 Defining Rot

We will use the notation $B(\mathcal{R})$ for $\bigcup_{R\in\mathcal{R}} B(R)$. Any immersion over $f$, restricts to one of $B(\mathcal{R})$ so we have the map:

$$[I] \longrightarrow [I(B(\mathcal{R}),\mathbb{R}^4)] \longrightarrow \underset{R\in\mathcal{R}}{\times} (H^1(R) \quad \mathbb{Z}) \ : \ [f,h] \longrightarrow (s_h, r_h).$$

In this paragraph we determine a necessary condition that an element of $\underset{R\in\mathcal{R}}{\times} \mathbb{Z} = \mathbb{Z}^{\mathcal{R}}$ be $r_h$ for some $(f,h) \in I$. To give this condition we will relate $r_h$, an orientation of $S(f)$, and a function $n : \mathcal{R} \times \mathcal{C} \longrightarrow \mathbb{Z}$ where $n(R,c)$ is intuitively the number of times $R$ abuts $c$. To define $n$ precisely, we recall (see §1.3) that if $c \in \mathcal{C}_0$, then the q-image of a transverse manifold at $x \in q^{-1}(c) \cap S(f)$ is just a closed interval with one endpoint on $c$ and the rest of the interval interior to one element $R \in \mathcal{R}$. On the other hand, if $c \in \mathcal{C}_1$, then the q-image of a transverse manifold to $x \in q^{-1}(c) \cap S(f)$ is a Y, where the arms of the Y are mapped by $\bar{f}$ to the same side of $\bar{f}(c)$ and the stem is mapped by $\bar{f}$ to the other side of $\bar{f}(c)$ locally.

Definition. For each $R \in \mathcal{R}$ and $c \in \mathcal{C}$, we let

$$n(R,c) = 0 \quad \text{if} \quad c \cap \partial R = \emptyset$$

$$n(R,c) = 1 \quad \text{if} \quad c \subseteq \partial R \quad \text{and} \quad c \in \mathcal{C}_0$$

$n(R,c) = A(R,c) - S(R,c)$ where $A(R,c) = 1$ or $2$ if $R$ contains one or both the arms of the Y across $c$ and $S(R,c) = 1$ or $0$ if $R$ does or doesn't contain the stem of Y across $c$.

An example in which $R$ contains all the "parts" of the Y across $c$ is: Here we have two elements $c_0, c_1 \in \mathcal{C}_1$ meeting at a vertex of type $(1\cdot2\cdot2\cdot3)$. A whole neighborhood of these two elements of $\mathcal{C}_1$ is contained in two elements of $\mathcal{R}$. Both arms of the Y across $c_0$ lie in $R_1$ and the stem is in $R_0$, but both arms and the stem of the Y across $c_1$ are in $R_1$. Thus $n(R_0,c_0) = -1$, $n(R_1,c_0) = 2$, $n(R_0,c_1) = 0$, $n(R_1,c_1) = 1$.

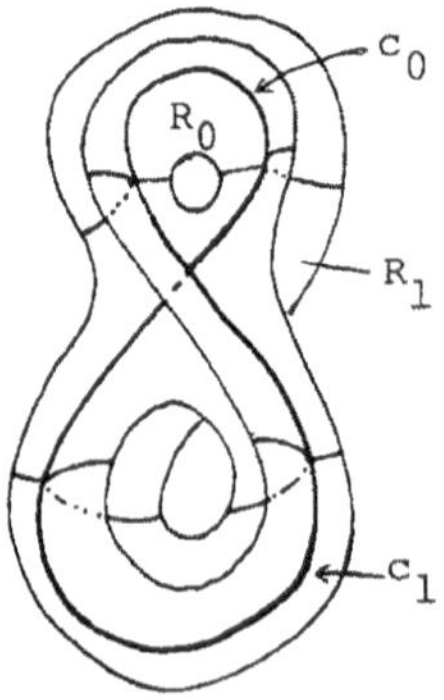

In §1.5 we introduced an orientation of the components of $\mathcal{C}$ which we call the <u>standard</u> <u>orientation</u>. Recall that this orientation of the components of $\mathcal{C}$ does not come from an orientation of $S(f)$ if there are any $(1\cdot2\cdot2\cdot1)$-points. An orientation of $S(f)$ induces an orientation of the components of $\mathcal{C}$. That is, an orientation of $S(f)$ gives a function $\vartheta : \mathcal{C} \longrightarrow \{\pm 1\}$ where $\vartheta(c) = +1\ (-1)$ if $q^{-1} \mid c$ is an orientation preserving (reversing) embedding of $c$ with its standard orientation into the oriented $S(f)$. By an abuse of notation, we will speak of $\vartheta$ as an <u>orientation</u> of $S(f)$ or of $q(S(f)) = \Sigma$, or of $\mathcal{C}$.

<u>Definition</u>. <u>Given</u> <u>an</u> <u>orientation</u> $\vartheta$ <u>of</u> $S(f)$, <u>let</u> $\mathrm{Rot}(\vartheta) \subseteq \mathbb{Z}^{\mathcal{R}}$ <u>such</u> <u>that</u> $r \in \mathrm{Rot}\ \vartheta$ <u>satisfies</u>:

$$\sum_{R \in \mathcal{R}} r(R)n(R,c) = \vartheta(c).$$

<u>We</u> <u>write</u> $\mathrm{Rot} = \bigcup_{\vartheta} \mathrm{Rot}(\vartheta)$.

<u>Theorem</u>. <u>If</u> $(f,h) \in I$ <u>then</u> $r_h \in \mathrm{Rot}$.

<u>Proof</u>. Let $(f,h) \in I$; we compute $\sum_R r_h(c)n(R,c)$ for each $c \in \mathcal{C}$. At $x \in S(f)$ above $c$, we have a transverse manifold $T = T(x)$ (see §1.3) to $x$. Since all $T(x)$ are diffeomorphic for $x \in c$, we speak of $T$ as a transverse manifold to $c$. $T$ is a disc if $c \in \mathcal{C}_0$ and is a disc with two holes if $c \in \mathcal{C}_1$. Our standard orientation of $c$ together with the orientation of $M$ specifies an orientation of $T$. The boundary of $T$ consists of oriented circles, each of which is a fibre of $q \mid B(R)$ for $c \subseteq \partial R$. If $s_i$ is a circle component of $\partial T$, let $R_i \in \mathcal{R}$ be such that $q(s_i) \in R_i$. We orient $s_i$ as fibres of $q \mid B(R_i)$; since $B(R_i) \subseteq M$ is oriented and $R_i$ is oriented by virtue of the immersion into $\mathbb{R}^2$, $\overline{f}$, we give the fibre of $q \mid B(R_i)$ an orientation consistent with these two orientations. Now $\partial T = \sum_i a_i s_i$, where $a_i = +1\ (-1)$ if $f(s_i)$ lies to the left (right) of the $f$-image of $c$ near $f(x)$. Since $\mathrm{rot}(h \mid s_i) = r_h(R_i)$, we have $\mathrm{rot}(h \mid \partial T) = \sum a_i r_h(R_i)$. If $c \in \mathcal{C}_0$, $\partial T$ has only one component and $\mathrm{rot}(h \mid \partial T) = r_h(R)$ for the unique $R$ abutting $c$, and $n(R,c( = 1$, so in this case $\mathrm{rot}(h \mid \partial T) = \sum_R n(R,c)r_h(R)$. If $c \in \mathcal{C}_1$, there are three component circles, two corresponding to the arms of the $Y = q(T)$ across $c$ and one corresponding to the stem. The corresponding $a_i$ are two $(+1)$'s and one $(-1)$. Once again we have $\mathrm{rot}(h \mid \partial T) = \sum_R n(R,c)r_h(R)$.

Lemma. $\text{rot}(h \mid \partial T) = (-1)^i$ (or $(-1)^{i+1}$) if $h \mid T$ is orientation preserving (or reversing) at $x \in q^{-1}(c) \cap S(f)$ for $c \in C_i$, $i = 0, 1$.

Proof. Note first that the statement makes sense since at each $x \in S(f)$, the tangent space to $T(x)$ at $x$ is the kernel of $Tf$ at $x$. Hence $h \mid T(x)$ is nonsingular at $x$.

If $c \in C_0$, this is obvious since $T$ is just a disc with $h$ nonsingular at its center, $x$. The non-degenerate fibres of $q$ in $T$ are just concentric circles in the disc, all immersed with the same rotation number, $\text{rot}(h \mid \partial T)$. Since $h$ is nonsingular at $x$, the center of the disc, a sufficiently small circle fibre near $x$ is embedded in $\mathbb{R}^2$ by $h$. Thus $\text{rot}(h \mid \partial T) = +1$ $(-1)$ according to whether $h \mid T$ preserves (or reverses) orientation at $x$.

If $c \in C_1$, $T$ is a disc with two holes. In the figure below we show $T$ with its $q$-fibres oriented as fibres of their respective $B(R)$. The convention that we use is that the arc of $S(f)$ which projects to $c$ points up out of the plane of the page. This as we've remarked determines the orientation of $T$. On all the level curves outside the figure eight, the rotation number of $h$ is constant, as it is on all of the pairs enclosed by the figure eight. Thus to compute $\text{rot}(h \mid \partial T)$ it suffices to examine how $h$ maps the boundary of a very thin neighborhood of the figure eight. We may assume that the only change that occurs in passing from the outside boundary circle to the two inside ones is that the image of:

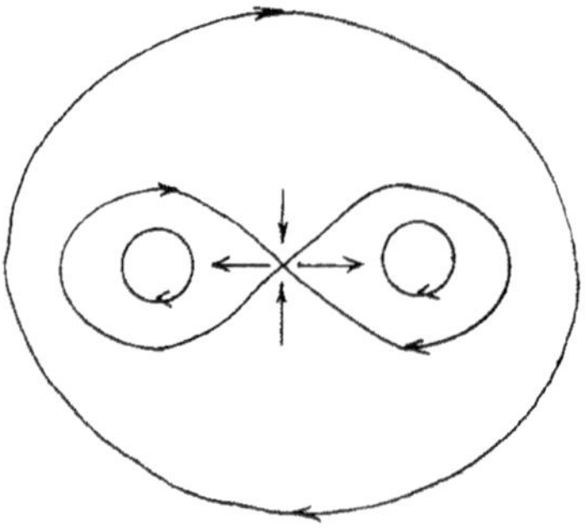

T with orientation 

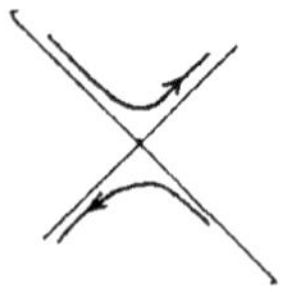 is replaced by 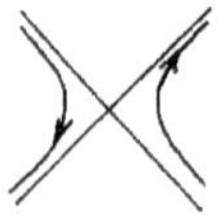

This gives a change in the rotation number of $(-1)$ if $h$ is orientation preserving and $(+1)$ if $h$ is orientation reversing. //

Thus we have found that $\sum_c \sum_R n(R,c) r_h(R) c = \sum_c \mathrm{rot}(h \mid \partial T) c = \sum_c (\pm 1) c$. We check that the $(\pm 1)$ coefficients give an orientation of $\Sigma$ that is induced by an orientation of $S(f)$. To do this we must check at each $v \in V$.

1) If $v$ is of type $(1\cdot 2\cdot 2\cdot 3)$, we have four arcs having $v$ as an end point oriented and labelled as shown below. The arcs $c_i$ and

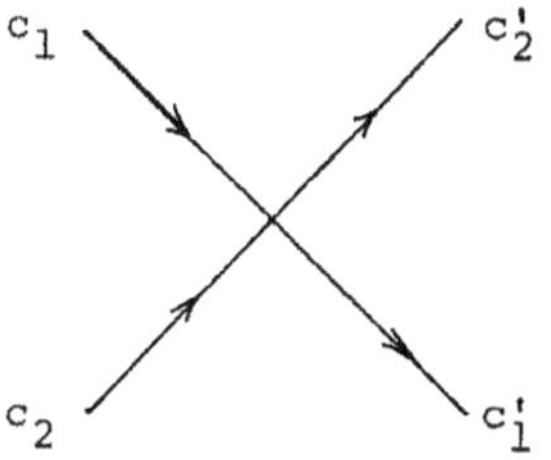


$c_i'$ are consistently oriented by the standard orientation. Thus the transverse manifolds to $c_i$ and $c_i'$ will be oriented the same way. Thus if $T_i$ and $T_i'$ are the transverse manifolds to $c_i$ and $c_i'$, then if $\varepsilon_i = \mathrm{rot}(h \mid T_i) = \mathrm{rot}(h \mid T_i')$, we see that in the sum $\sum n(R,c) r_h(R) c$, these four arcs will appear as $\varepsilon_1(c_1 + c_1') + \varepsilon_2(c_2 + c_2')$, $|\varepsilon_i| = 1$. This part of the oriented chain obviously comes from an orientation of the part of $S(f)$ that projects to this part of $\Sigma$.

2) If $v$ is of type $(1\cdot 2\cdot 2\cdot 1)$, we have four arcs as shown in the figure. If $T_i$ and $T_i'$ are the transverse manifolds to $c_i$ and

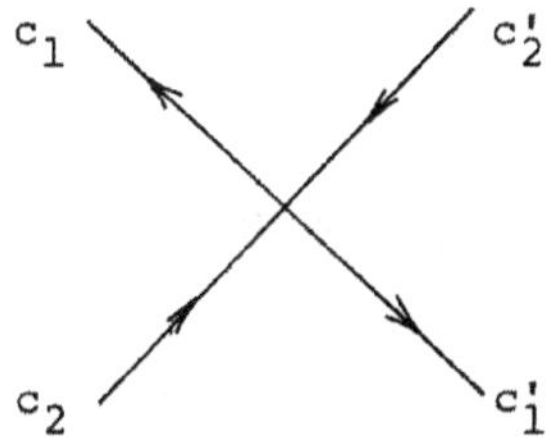


$c_i'$ respectively we have $\mathrm{rot}(h \mid T_i)$ $\mathrm{rot}(h \mid T_i') = -1$. So these arcs appear in the chain as $\varepsilon_1(c_1 - c_1') + \varepsilon_2(c_2 - c_2')$. Again this orientation is induced by an orientation of $S(f)$ which projects to this part of $\Sigma$.

3) If $v$ is a cusp. We have two arcs $c_1 \in C_1$ and $c_0 \in C_0$ having $v$ as an end point. By working close enough to the cusp, the

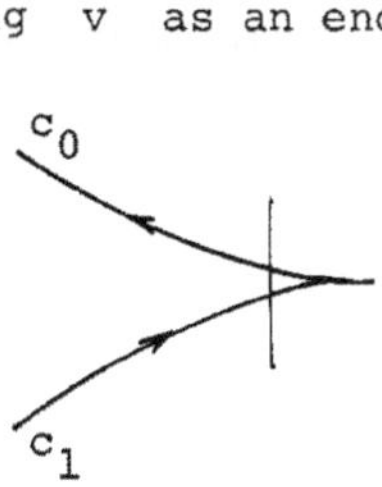


transverse manifolds to both $c_1$ and $c_0$ may be taken as contained in the $q$ preimage of an arc transverse to both $c_1$ and $c_0$. The orientations of the transverse manifolds $T_i$ to $c_i$ induce opposite orientations to this containing manifold. If $x_i \in q^{-1}(c_i) \cap S(f)$, then one of $h \mid T_i$ at $x_i$ is orientation preserving and the other is reversing. Thus by the lemma, $\mathrm{rot}(h \mid T_0)\,\mathrm{rot}(h \mid T_1) = 1$. Thus these arcs appear in our chain as $\varepsilon(c_1 + c_0)$ as required. //

Given an orientation $\vartheta$ of $S(f)$, the set $\mathrm{Rot}(\vartheta)$ may be large; it is the set of integral solutions $r \in \mathbb{Z}^{R}$ of the system of $|C|$ linear equations in $|R|$ unknowns:

$$(*_{\vartheta}) \qquad \sum_{R \in R} r(R)n(R,c) = \vartheta(c).$$

However there is a unique function analogously defined on the components of $\mathbb{R}^2 - f(S(f))$. This is the content of the following proposition. It is included here as an aside and will not be used in the sequel.

Proposition. *Let* $\vartheta$ *be any orientation of* $S(f)$, *and let* $f(S(f))$ *have the induced orientation. Let* $\mathcal{D}$ *be the set of components of* $\mathbb{R}^2 - f(S(f))$.

a) *There is a unique function* $\sigma : \mathcal{D} \longrightarrow \mathbb{Z}$ *such that*

(0) *if* $D \in \mathcal{D}$ *and* $D \,||\, f(M)$, *then* $\sigma(D) = 0$

(1) *if* $D, D' \in \mathcal{D}$ *have a common boundary arc, say* $\alpha \subseteq f(S(f))$, *then* $\sigma(D') = \sigma(D) + 1$ if $D'$ *lies to the left and* $D$ *lies to the right of* $\alpha$, *where* $\alpha$ *is oriented by* $\vartheta$.

b) *If* $r \in \mathrm{Rot}(\vartheta)$, *then* $\sigma(D) = \sum m(R,D)r(R)$ *where the sum is over all* $R$ *for which* $\bar{f}(R) \cap D \neq \emptyset$, $m(R,D)$ *is the number of points in* $R$ *in the* $\bar{f}$*-preimage of a point in* $D$.

To see that $m(R,D)$ is well defined argue as follows: $\bar{f} : R \longrightarrow \mathbb{R}^2$ is an immersion, and $\mathbb{R}^2 - \bar{f}(\partial R)$ is a union of a finite number of open connected sets. If $D$ is one of those sets, let $D_m = \{y \in D \mid \bar{f}^{-1}(y) \cap R$ has $m$ elements$\}$. These sets $D_m$ are open in $D$. Hence, $D_m = D$ for some $m$. This integer is our $m(R,D)$.

Proof. a) If $S(f)$ has only one component embedded in $\mathbb{R}^2$ by $f$, then $\sigma$ is obvious. If $S(f)$ consists of finitely many components each of which is embedded by $f$, $\sigma$ is just the sum of the $\sigma$'s for each component. To reduce to the embedded case, when components have normal crossings, just replace  by . If the values of $\sigma$ for this new configuration are 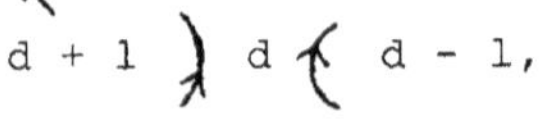 then the values of $\sigma$ in the original configurations are 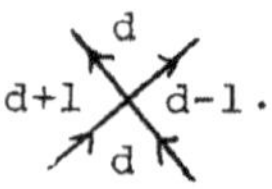. Here the existence of cusps is irrelevant since wherever "embedding" occurs in this proof we may substitute "embedding with cusps" (i.e. a homeomorphism into, that is an embedding except for cusps).

b) Suppose $D$ and $D'$ have a common arc, $\alpha$ in their boundaries. There is an element $c \in C$ such that $\overline{f}(\vartheta(c)c) \supseteq \alpha$ as oriented arcs. We must prove: $\sum_R m(R,D')r(R) = \sum_R m(R,D)r(R) + 1$, if $D'$ is on the left and $D$ on the right of $\alpha$. If $y \in \alpha$, the number of points in $\overline{f}^{-1}(y) \cap R^0$ is independent of the choice of $y \in \alpha$. Call this integer $k(R,\alpha)$. Thus it is enough to prove that

$$\sum_R \{(m(R,D') - k(R,\alpha)) - (m(R,D) - k(R,\alpha))\}r(R) = 1$$

for $D'$ to the left of and $D$ to the right of $\alpha$. However the expression in brackets is just $\vartheta(c)n(R,c)$, since we are counting the number of times $R$ abuts $c$ so that its image lies to the left of $\alpha$ less the number of times $R$ abuts $c$ so that its image lies to the right of $\alpha$. But the resulting equation:

$$\sum_R \vartheta(c)n(R,c)r(R) = 1$$

holds since $r \in \mathrm{Rot}(\vartheta)$. //

It is easy to see how the non-uniqueness of the solutions to $(*_\vartheta)$ comes about in spite of the uniqueness of the above defined $\sigma$. If we had three regions, $R_s$, $R$, $R'$ abutting an element $c \in C_1$, where the $\overline{f}$ images of $R$ and $R'$ are on one side of the $\overline{f}$-image of $c$ and the $\overline{f}$-image of $R_s$ is on the other side. Then $(*_\vartheta)$ requires that $r$ satisfy:

$$r(R) + r(R') - r(R_s) = \vartheta(c).$$

So that even specifying $r(R_s) = a$, given any $b \in \mathbb{Z}$, $(*_\vartheta)$ allows $r(R) = a + b + \vartheta(c)$ and $r(R') = -b$.

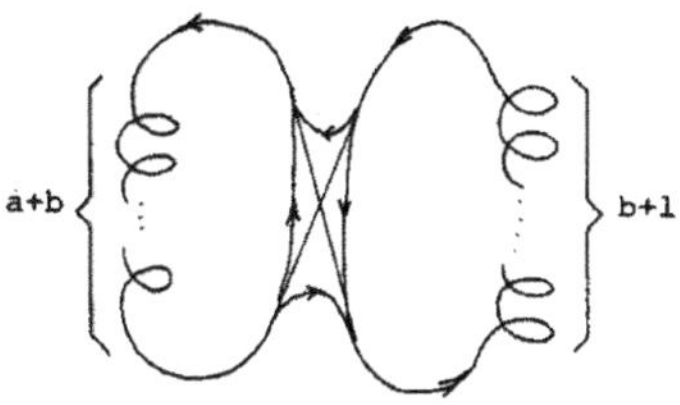


Transition from rotation number
a to (a+b+1) and (-b)

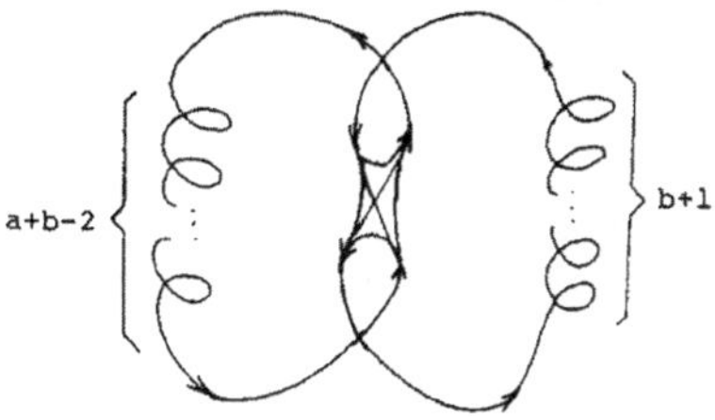


Transition from rotation number
a to (a+b-1) and (-b)

To see that a given map $f : M \longrightarrow \mathbb{R}^2$ and a given orientation ϑ of $S(f)$ can admit infinitely many distinct lifts corresponding to different choices of $r$, let $M = \mathbb{T} \times \mathbb{S}^1$ where $\mathbb{T} = \mathbb{S}^1 \times \mathbb{S}^1$ and $f : \mathbb{T} \times \mathbb{S}^1 \longrightarrow I \times \mathbb{S}^1 \subseteq \mathbb{R}^2$, is the product of the standard height function on the torus and the identity map of $\mathbb{S}^1$ with itself. Restricted to each torus our map is shown in the figure and $W$ is the product of  with a circle namely:

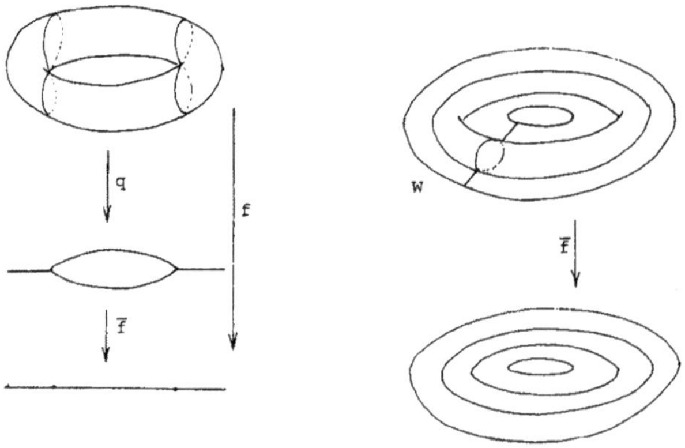


The three figures give the $r$-values in parentheses for the four regions of $W$ for the given orientation of $\Sigma$.

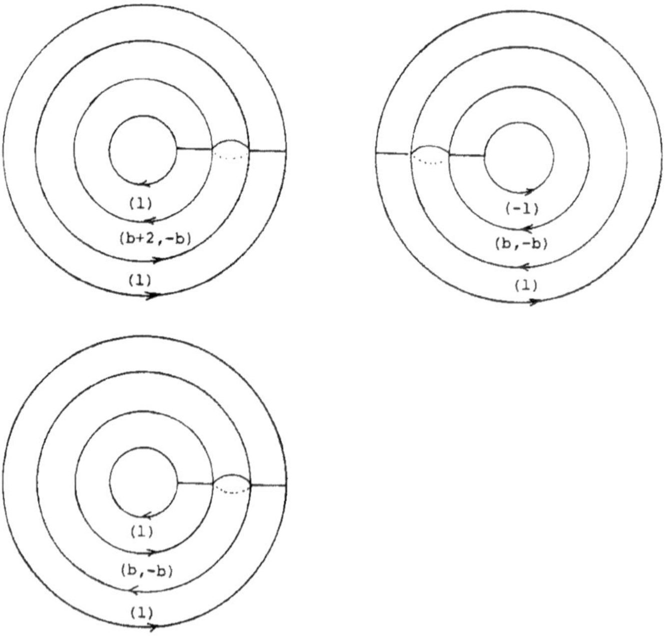

These maps easily lift to $\mathbb{R}^4$ and all represent different f-regular homotopy classes. Except for reversing all the orientation of $\Sigma$ which would result in changing all of the signs of the r-values, these are all of the possible r-functions for the map f. However this does not exhaust the list of f-regular homotopy classes of lifts of f, since for each region R, $H_1(R,\mathbb{Z}) = \mathbb{Z}$ and a lift can also spin the immersed circle fibre over a point of R as the point travels around a generator of $H_1(R,\mathbb{Z})$. This is what is measured by $s_h$ for $(f,h) \in I$. (See §3.1 for the definition of $s_h$.)

We end this section with the description of two stable maps that cannot be lifted to immersions into $\mathbb{R}^4$. In the first M is not orientable and in the second M is orientable.

1) Let M be the identification space of $I \times S^2$ where we identify $0 \times S^2$ with $1 \times S^2$ by reversing the orientation of $S^2$. This resulting manifold is an $S^2$ bundle over the circle. We map this manifold to the annulus $S^1 \times [-1,1]$ in $\mathbb{R}^2$ by mapping each $S^2$ to the radial interval of the annulus. The orientation reversing diffeomorphism used in the identification is chosen to commute with this map (i.e. if $S^2 \subseteq \mathbb{R}^3 = \{(x,y,z) \in \mathbb{R}^3 \mid x^2 + y^2 + z^2 = 1\}$, the map of $S^2 \longrightarrow [-1,1]$ is the restriction of the projection

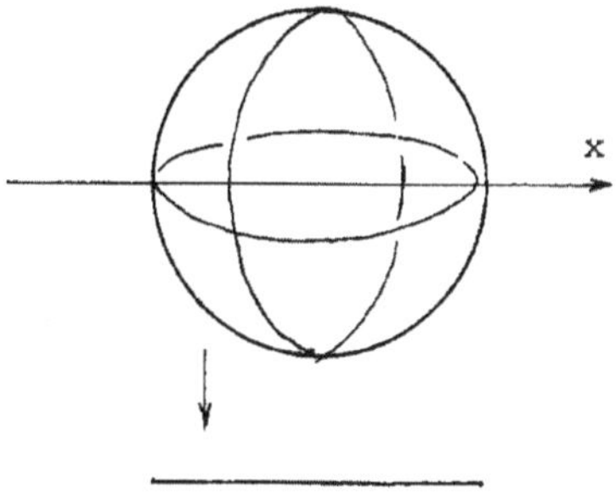


$(x,y,z) \longrightarrow x$. The orientation reversing diffeomorphism is the restriction of $(x,y,z) \longrightarrow (x,-y,z)$.) By construction the f-preimage of any circle interior to the annulus, $S^1 \times 0 \subseteq \mathbb{R}^2$, is a Klein bottle in M. Thus by the note at the beginning of §3.1, if $(f,h)$ is an immersion of M in $\mathbb{R}^4$, h must immerse the circle fibre over any point of $S^1 \times 0$ with rotation number, zero. However $W_f = S^1 \times [-1,1]$, and $\mathcal{R}$ consists of a single element, namely $S^1 \times (-1,1) = R$. Thus the only possible value of $r(R)$ is zero. On the other hand, the two singular curves which project down to the boundary curves of the annulus $S^1 \times \{\pm 1\}$ are both definite so that the f-fibre circles above points near $S^1 \times \{\pm 1\}$ must be embedded by h for any lift $(f,h)$ of f. Hence $|r(R)|$ must be 1. Contradiction. Thus $I = \emptyset$.

2) In this example we do not describe M explicitly but we describe W and f(M). In this case, I don't know a convenient description of M; W gives a pattern for its construction.

We suppose that $f(M) \subseteq \mathbb{R}^2$ looks like the figure below: The numbers in the regions indicate the number of times $W$ covers the region. Further assume that the $\bar{f}$-preimage of the double covered region is connected. The three embedded circles are images of elements of $C_0$, the circle in the region labelled $\bar{f}(R_2)$ being doubly covered. The immersed circle is the image of two elements of $C_1$ and the self intersection point is the image of a point of type $(1 \cdot 2 \cdot 2 \cdot 1)$. $R$ consists of three elements $R_1$, $R_2$, $R_3$. We show that for this configuration, $\mathrm{Rot} = \emptyset$ so $I = \emptyset$ as well. Suppose $r \in \mathrm{Rot}$. Let $r_i = r(R_i)$. Since every region $R_i \in R$ has an element of $C_0$ as part of its boundary, $|r_i| = 1$ for all $i$. For any orientation, $\vartheta$, of the four curves $(*_\vartheta)$ gives in addition:

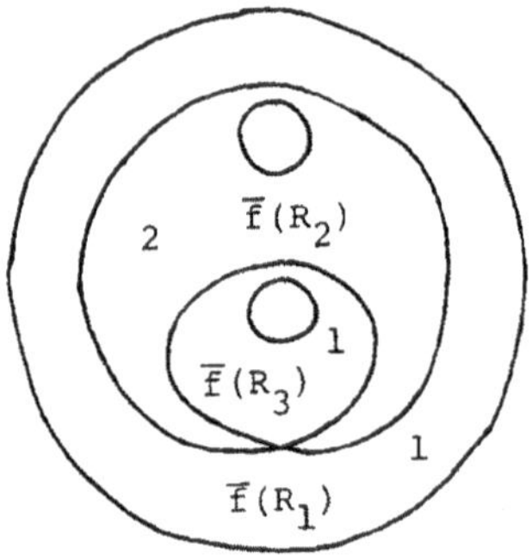


$$2r_2 - r_1 = \vartheta(c_1), \quad 2r_2 - r_3 = \vartheta(c_2),$$

where $c_1$ is the element of $C_1$ separating $R_1$ and $R_2$ and $c_2$ is the element of $C_1$ separating $R_2$ from $R_3$. Obviously $\vartheta(c_1) = -\vartheta(c_2)$, so we have

$$4r_2 = r_1 + r_3$$

which has no solution with $|r_i| = 1$, all $i$.

## 3.3 <u>Fixing the immersion of</u> $S(f)$ <u>over</u> $f \mid S(f)$, $[I] = [I_E]$

Denote by $f|$, $f \mid S(f)$.

<u>Proposition</u>. <u>Let</u> $E = (f|, e) : S(f) \longrightarrow \mathbb{R}^2 \times \mathbb{R}^2$ <u>be any embedding. In every</u> f-<u>regular homotopy class of immersions of</u> $M$ <u>in</u> $\mathbb{R}^4$, <u>there is a representative whose restriction to</u> $S(f)$ <u>is</u> $E$.

Thus if we define $I_E \subseteq I$ as the set of all $(f,h)$ such that $(f,h) \mid S(f) = E$, the proposition states that $[I_E] = [I]$. N.B. that $[I_E]$ are f-regular homotopy classes of elements in $I_E$; the homotopies are <u>not</u> assumed constantly $E$ on $S(f)$.

Proof. Given any immersion $(f,h) \in I$, it is easy to find $(f,h') \in [f,h]$ so that $(f,h') \mid S(f)$ is an embedding. In particular, we need merely separate a finite number of double points on a finite union of immersed circles in $\mathbb{R}^4$.

We will show: Given $E_0 = (f|, e_0)$ and $E_1 = (f|, e_1)$ two embeddings of $S(f)$ in $\mathbb{R}^4$, there is an isotopy of the identity map of $\mathbb{R}^2 \times \mathbb{R}^2$ of the form $(1\times\phi_t)$ such that $(1\times\phi_1)\circ E_0 = E_1$. This obviously gives the proposition since if $E_0 = (f,h) \mid S(f)$, then $(1\times\phi_1)\circ (f,h) = (f,h') \in [f,h]$ and $(f,h') \mid S(f) = E_1$.

We construct the isotopy in three stages. The first of which moves $E_0$ to coincide with $E_1$ on the double points of $f$, the second of which does the same for the cusp points of $f$, the third stage takes care of the rest.

(i) If $f(z_1) = f(z_2) = x_0$, $e_j(z_1) = y_{j1} \neq y_{j2} = e_j(z_2)$ for $j = 0, 1$. Let $v_j = \frac{1}{2}(y_{j1}+y_{j2})$. Choose $\theta \in [-\pi,\pi]$ such that

$$e^{i\theta}\frac{(y_{01}-v_0)}{|y_{01}-v_0|} = \frac{y_{11}-v_1}{|y_{11}-v_1|}, \quad \text{(here we identify } \mathbb{R}^2 \text{ with } \mathbb{C}).$$

Let $\lambda : \mathbb{R}^2 \longrightarrow [0,1]$ be a bump function at $x_0$ i.e. outside a disc, $D$, about $x_0$, $\lambda = 0$ and on a neighborhood of $x_0$ in $D$, $\lambda = 1$. Define $\phi_t : \mathbb{R}^2 \times \mathbb{R}^2 \longrightarrow \mathbb{R}^2$ by:

$$\phi_t(x,y) = (1-t\lambda(x)y+t\lambda(x)\left[v_1 + \frac{|y_{11}-v_1|}{|y_{01}-v_0|}e^{it\lambda(x)\theta}(y-v_0)\right].$$

To check that $1 \times \phi_t$ is a diffeomorphism for each $t \in [0,1]$ we examine the coefficient of $y$ in $\phi_t(x,y)$:

$$\left[1 - t\lambda(x)\right] + t\lambda(x)\left[\frac{|y_{11}-v_1|}{|y_{01}-v_0|}e^{it\lambda(x)\theta}\right].$$

This cannot vanish unless $e^{it\lambda(x)\theta} = -1$, in which case $t\lambda(x) = 1$. In that case the coefficient is $-\dfrac{|y_{11}-v_1|}{|y_{01}-v_0|} \neq 0$. To see that $(1\times\phi_1)\circ E_0(z_j) = E_1(z_j)$ we easily compute:

$$\phi_1(x_0,y_{0j}) = v_1 + \frac{|y_{11}-v_1|}{|y_{01}-v_0|}e^{i\theta}(y_{0j}-v_0) = y_{1j}.$$

Choose disjoint discs about all the $f$-images of double points of $f$ and construct isotopies of the identity as above for each one.

Composing this finite collection of isotopies produces an isotopy of the identity of the form $1 \times \Phi_t$ such that $(1\times\Phi_1)\circ E_0$ agrees with $E_1$ on the double points of $f|$.

(ii) Let $z_0$ be a cusp point of $f$ and $x_0 = f(z_0)$. Let $e_j(z_0) = y_j$ and $(de_j/dz)(z_0) = e_j'(z_0) = y_j' \neq 0$. Choose $\theta \in [-\pi,\pi]$ so that $e^{i\theta}(y'/|y_0'|) = y_1'/|y_1'|$. Let $\lambda : \mathbb{R}^2 \longrightarrow [0,1]$ be a bump function at $x_0$. Define $\phi_t : \mathbb{R}^2 \times \mathbb{R}^2 \longrightarrow \mathbb{R}^2$ by:

$$\phi_t(x,y) = (1-t\lambda(x))y + t\lambda(x)\left[y_1 + (|y_1'|/|y_0'|)e^{it\lambda(x)\theta}(y-y_0)\right].$$

Once again the coefficient of $y$ is

$$(1-t\lambda(x)) + t\lambda(x)\left\{(|y_1'|/|y_0'|)e^{it\lambda(x)\theta}\right\}$$

which is never zero.

Finally $(1\times\phi_1)\circ E_0(z_0) = x_0, \phi_1(x_0,y_0) = (x_0,y_1) = E_1(z_0)$, and $(1\times\phi_1\circ E_0)'(z_0) = (0,(|y_1'|/|y_0'|)e^{i\theta}e_0'(z_0) = (0,y_1') = E_1'(z_0)$.

As in (i) we can construct an isotopy of the identity $(1\times\psi_t)$ of $\mathbb{R}^2 \times \mathbb{R}^2$ so that at every cusp point of $f$, the values and the tangents of $(1\times\psi_1)\circ E_0$ and $E_1$ agree.

(iii) Composing the isotopies of parts (i) and (ii) we may assume that $E_0$ and $E_1$ are embeddings of $S(f)$ over $f|$ which agree on the double points and cusps of $f|$ and whose tangent maps at the cusp points of $f$ also coincide. Consider the homotopy of maps over $f|$, $E_t = tE_1 + (1-t)E_0 = (f|,e_t)$. It is obvious that $E_t$ is an embedding for all $t$; it is an immersion since at any cusp, $z$, $E_t'(z) = E_0'(z) = E_1'(z) \neq 0$, and it is 1:1 since $E_t = E_0 = E_1$ at the double points of $f$.

It is easy to construct an isotopy of the identity of the form $(1\times\upsilon_t)$ such that $(1\times\upsilon_1)\circ E_0 = E_1$. This is a special case of the result in [Lima, Palais], the special form of the isotopy being a consequence of the constancy in $t$ of the first $\mathbb{R}^2$-component of $E_t$. //

We now choose a specific embedding $E$ of $S(f)$ over $f \mid S(f)$ in $\mathbb{R}^4$ which will be fixed from now on. In particular our embedding will carry $S(f)$ into $\mathbb{R}^2 \times \mathbb{R} \times 0 \subseteq \mathbb{R}^4$. Except for neighborhoods of double points and cusps the image, $E(z)$ is merely $(f(z),0)$. If $f(z_1) = f(z_2)$, $z_i \in S(f)$, we define $E$ near these $z_i$ by spreading the image of $S(f)$ symmetrically into $\mathbb{R}^2 \times \mathbb{R} \times 0$. We embed a neighborhood of a cusp symmetrically above and below the $\mathbb{R}^2 \times 0$ plane in $\mathbb{R}^2 \times \mathbb{R} \times 0$. We illustrate:

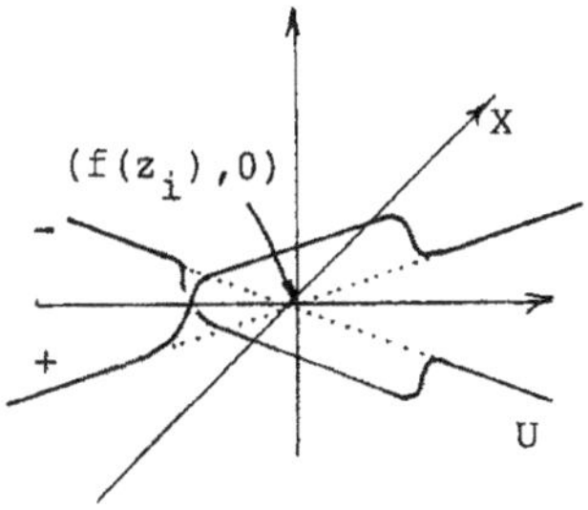


E near double points
$z_1$ and $z_2$ of $f|$

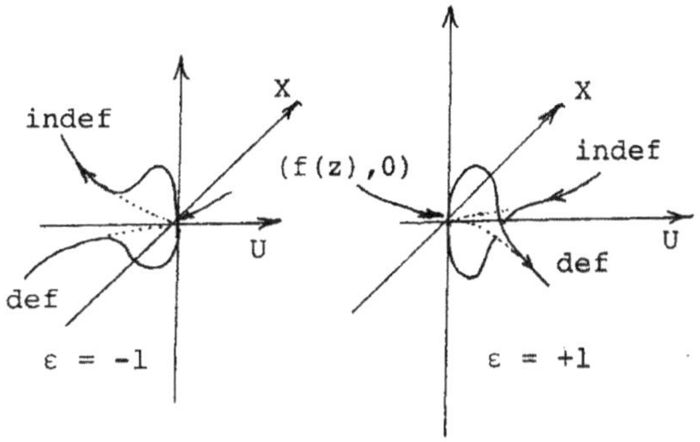


E near an ε-cusp

The indicated $(U,X)$ coordinates are those of $I \times J$ in the $C^2$PN of $f$ at a vertex.

In the next section, when we standardize the germs of our immersions at the vertices, $V$, this embedding will be a bit more precisely specified.

It is easy to see using the preceding proposition that

<u>Corollary</u>. <u>If</u> $G = (f|,g) : S(f) \longrightarrow \mathbb{R}^2 \times \mathbb{R}^2$ <u>is any immersion, then in every</u> f-<u>regular homotopy class of immersion of</u> $M$ <u>in</u> $\mathbb{R}^4$ <u>there is a representative whose restriction to</u> $S(f)$ <u>is</u> $G$.

In the notation introduced: $[I_G] = [I]$.

In most of what follows we will use the embedding $E$ of $S(f)$ in $\mathbb{R}^4$. That $E$ is an embedding is a convenience in visualizing the maps, $h$ onto $\mathbb{R}^2$ of the transverse manifolds through non-simple vertices but it is inessential. The only place that the fact that $E$ separates double points is used is in the explicit formulas for the germs at those points in the definition of $I_V$. If instead of $E$ the above-described embedding we take $E_i$ to be the immersion that agrees with $E$ in the neighborhoods of the cusps but is otherwise defined by $e_i = 0$, the changes that are required in what follows are obvious and are indicated where they come up in §3.4 and 3.5. Rather than carry both possibilities along, we will work with the embedding $E$ until §3.10 thereafter we work with the immersion $E_i$ where it is more convenient.

3.4 Standard forms for germs of lifts at the vertices, V; $[I_{V_*}]_E = [I_E]_E$, $[I_{V_*}] = [I_E]$.

As described in §1.5 all of the arcs of $C$ are oriented. Since $M$ is also oriented we have a canonical orientation for any transverse manifold at any point of $c \in C$.

Definition. For any immersion $(f,h) \in I$, we define $\delta_h : C \longrightarrow \{+1,-1\}$ by $\delta_h(c) = +1$ $(-1)$ if the germ of $h$ restricted to a transverse manifold to $c$, at any point of $c$ is orientation preserving (reversing). It is easy to see that:

(i) If $c_1$ and $c_2$ are in $C_1$ and are arcs of the same component of $S(f)$, then $\delta_h(c_1) = \delta_h(c_2)$, $(= -\delta_h(c_2))$ if the standard orientations of $c_1$ and $c_2$ are compatible (incompatible).

(ii) If $c_1$ and $c_2$ are arcs of the same component of $S(f)$ which meet at a cusp then $\delta_h(c_1) = -\delta_h(c_2)$.

Let $\mathcal{D}$ be the set of maps $(C,\{-1,+1\})$ which satisfy (i) and (ii). Obviously $\delta_h$ is constant on f-fibre regular homotopy classes so we have a well defined map:

$$\delta : [I] \longrightarrow \mathcal{D} : [f,h] \longrightarrow \delta_h.$$

In this section we show that in every f-regular homotopy class, $[f,h]$, there is a representative whose germs at each point of $q^{-1}(V) \cap S(f)$ are canonically given--depending only on $\delta_h$.

For any $\delta \in \mathcal{D}$ we extend its domain to the points of $q^{-1}(V) \cap S(f)$ in a somewhat arbitrary way:

If $v$ is a cusp, $\delta(v) = \delta(c)$ where $c$ is the element of $C_0$ abutting $v$.

Let $v$ be a non-simple vertex and let $(\phi_v, \psi_v, g_v)$ be our $C^2$PN of $f$ at $v$. The set $q^{-1}(v) \cap S(f)$ consists of two points $v^+ = \phi_v(P^+)$ and $v^- = \phi_v(P^-)$. The two arcs of the component of $S(f)$ that abut $v^+$ are $c^+$ and $c'^+$ and those that abut $v^-$ are $c^-$ and $c'^-$.

If $v$ is of type $(1\cdot2\cdot2\cdot1)$ the arcs in these pairs $(c^+,c'^+)$ and $(c^-,c'^-)$ are inconsistently oriented. These arcs are so labelled that $\phi_v \mid (I-\{0\}) \times \{P^-,P^+\}$ is orientation preserving on $\phi_v^{-1}(c^+)$ and on $\phi_v^{-1}(c'^-)$ (see §2.2 and the figure below). The orientations of $c^+$ and $c'^-$ give the same orientation to the transverse manifold at $v$. We define:

$$\delta(v^+) = \delta(c^+) = -\delta(c'^+)$$

and

$$\delta(v^-) = \delta(c'^-) = -\delta(c^-).$$

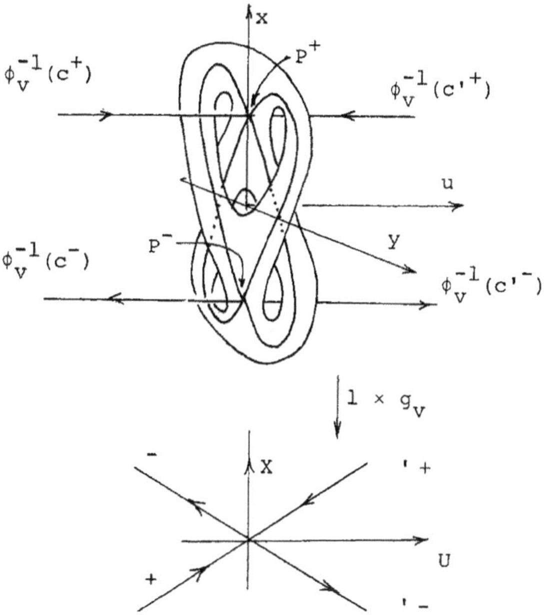


If $v$ is of type $(1\cdot2\cdot2\cdot3)$, the two arcs of $S(f)$ through $v^+$ and $v^-$ are consistently oriented and in our canonical coordinatized product neighborhoods $\phi_v \mid I - \{0\} \times \{P^+, P^-\}$ is orientation preserving onto $c^{\pm}$ and $c'^{\pm}$ (see §2.2 and the figure below).

We define

$$\delta(v^+) = \delta(c^+) = \delta(c'^+),$$

and

$$\delta(v^-) = \delta(c^-) = \delta(c'^-).$$

We use this function $\delta$ to describe a set of immersions over $f$ which are particularly simple near $q^{-1}(V) \cap S(f)$.

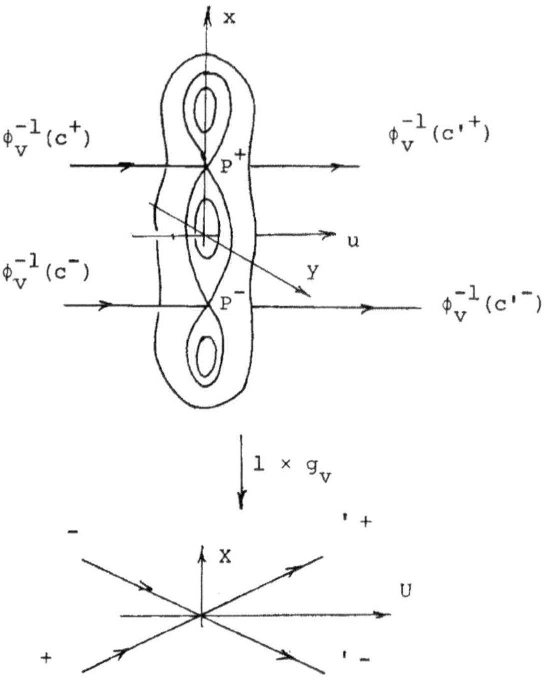


Recall that $I_E$ is the set of all immersions $(f,h)$ of $M$ in $\mathbb{R}^4$ such that $(f,h) \mid S(f) = E$. For each vertex $v$, define $k_v = h \circ \phi_v$ where $(f,h) \in I_E$ and $(\phi_v, \psi_v, g_v)$ is canonical. For all $(f,h) \in I_E$, $k_v \mid S(1 \times g_v)$ are the same. In particular, if $v$ is an $\varepsilon$-cusp, $k_v(3\varepsilon x^2, x, 0) = (x,0)$ and if $v$ is a non-simple point, $S(1 \times g_v) \cap (I \times \mathbb{D}_\pm) = I \times \{(\pm x_0, 0)\}$ and $k_v(u, \pm x_0, 0) = (\pm x_0, 0)$. (See §3.3 and §2.2.) In §3.3 we described the map $h \mid S(f)$ near a non-simple vertex as giving rise to $k_v(u, \pm x_0, 0) = (\pm \text{constant}, 0)$. Here we have merely chosen a convenient constant.

For each closed component $c \in C$, let $(\phi_c, \psi_c, g_c)$ be a $C^2$PN of $f$ along $c$ (see §2.3). Let $Q = (\pm(1-\alpha/2), 0) \in I_\pm \times T(c)$. We denote the common value $\phi_c(Q_+) = \phi_c(Q_-)$ by $v^c$ which we may assume without loss of generality is not a double point of $f \mid S(f)$. On $I_\pm \times T(c)$, for any $(f,h) \in I_E$, we define $k_{c,\pm} = h \circ \phi_{c,\pm}$ where $\phi_{c,\pm} = \phi_c \mid I_\pm \times T(c)$. Since $S(1 \times g_c) = I \times \{0\} \subseteq I \times T(c)$ and $h$ maps a neighborhood of $v^c$ in $S(f)$ to $0$ in $\mathbb{R}^2$, we may assume that $k_{c,\pm}(u,(0,0)) = 0$ for $u \in I_\pm$.

For each closed component $c \in C$, let $q(v^c) = v_c$ and let $V_* = V \cup \{v_c \mid \text{closed } c \in C\}$, and $V^* = q^{-1}(V_*) \cap S(f)$. We extend the definition of $\delta$ to all of $V^*$ by setting $\delta(v^c) = \delta(c)$ for closed $c \in C$.

<u>Definition of</u> $I_{V_*}$. An element $(f,h) \in I_E$ is in $I_{V_*}$ if the germs of $k_v$ at each point of $\phi_v^{-1}(q^{-1}(v) \cap S(f))$ for each vertex $v$, and the germs of $k_{c,\pm}$ at each point of $\phi_{c,\pm}^{-1}(v^c)$ for each closed component $c \in C$ are given by:

1) If $v$ is a cusp, $k_v(u,x,y) = (x, \delta_h(v)y)$.

2) If $v$ is a non-simple vertex, $k_{v^\pm}(u,x,y) = (x, \delta_h(v^\pm)y)$ where $k_{v^\pm}$ is the germ of $k_{v^\pm}$ at $\phi_v^{-1}(v^\pm) = P^\pm$.

3) For $c \in C$ a closed component, since $k_{c,+} = h \circ \phi_{c,+} = h \circ \phi_{c,-} \circ \phi_{c,-}^{-1} \circ \phi_{c,+} = k_{c,-} \circ \phi_{c,-}^{-1} \circ \phi_{c,+}$, to specify $k_{c,\pm}$ it suffices to specify $k_{c,+}$. We require that $k_{c,+}(u,x,y) = (x, \delta_h(v^c)y)$. (Recall that $\phi_{c,-}^{-1} \circ \phi_{c,+}(u,x,y) = (u+\alpha-2, \pm(x,y))$, see Proposition 2 (for $c$ closed) of §2.3.)

Let $I_V \supseteq I_{V_*}$ be the set of $(f,h)$ whose germs at $\hat{V} = q^{-1}(V) \cap S(f)$ are fixed by 1), 2) above.

<u>Note</u>. If $E_i$ were used instead of $E$, in the above definitions, 2) would be replaced by

$2_i$) ..., $k_{v^\pm}(u,x,y) = (x \mp x_0, \delta_h(v^\pm)y)$.

<u>Notation</u>. For any subset $I_*$ of $I_E$, let $[I_*]_E$ be the set of f-regular homotopy classes of maps in $I_*$ which are constantly $E$ when restricted to $S(f)$ during the homotopy.

<u>Proposition</u>. <u>The inclusions</u> $[I_{V_*}] \longrightarrow [I_V] \longrightarrow [I_E]$ <u>and</u> $[I_{V_*}]_E \longrightarrow [I_{V_*}]_E \longrightarrow [I_E]_E$ <u>are 1-1 correspondences</u>.

<u>Proof</u>. Since $f$ has rank one at every point of $S(f)$, we know that if $(f,h) \in I$ then $h$ has rank two at every point of $S(f)$. Thus we may take the germs: $1 \times k_v$ at each point of $\phi_v^{-1}(q^{-1}(v) \cap S(f))$ in $I \times T(v)$ for each vertex $v \in V$ and $1 \times k_{c,\pm}$ at each point of $\phi_{c,\pm}^{-1}(v^c)$ in $I_\pm \times T(c)$ for each closed element $c \in C$ to be germs of embeddings. Thus it is enough to prove:

<u>Lemma</u>. <u>Let</u> $1 \times k : I \times \mathbb{D} \longrightarrow I \times \mathbb{R}^2$ <u>be an embedding with either</u>

1) $k(u,(0,0)) = 0$

2) $k(3\varepsilon x^2, (x,0)) = (x,0)$.

<u>Then there is a regular homotopy</u> $1 \times k_t$ <u>with</u> $k_0 = k$ <u>and</u>

(i) $k_t$ <u>is independent of</u> $t$ <u>in a neighborhood of the boundary</u> $I \times \mathbb{D}$.

(ii) <u>On a sufficiently small neighborhood</u> $(I' \times \mathbb{D}')$ <u>of</u> $(0,0)$, $k_1(u,x,y) = (x,y)$, (<u>or</u> $= (x,-y)$) <u>if</u> $1 \times k$ <u>is orientation preserving</u> (<u>or reversing</u>).

(iii) <u>If</u> $V$ <u>is an open set in</u> $\mathbb{D}$ <u>and</u> $u_0 \in I$ <u>such that</u> $k(u_0,(x,y)) = (x,\pm y)$ <u>for</u> $(x,y) \in V$, <u>then</u> $k_t(u_0,(x,y)) = (x,\pm y)$ <u>for all</u> $t$ <u>and all</u> $(x,y) \in V$.

(iv) <u>Under hypothesis</u>: <u>1</u>) $k_t(u,(0,0)) = 0$, <u>and 2</u>) $k_t(3\varepsilon x^2,(x,0)) = (x,0)$ <u>for all</u> $t$.

The proof of this Lemma is analogous to, but simpler than that of Lemma 0 of §2.1. The details appear in the appendix.

## 3.5 <u>Standard forms for germs of lifts at</u> $S(f)$; $[I_S]_S = [I_V]_V$, $[I_S]_E = [I_V]_E$, $[I_S] = [I_V]$

In this paragraph we define standard forms for the germs of $(f,h) \in I$ along $S(f)$ and show that in every class of $[I] = [I_E] = [I_V]$ there is an element whose germ along $S(f)$ is standard. For such a "standardized-along-$S(f)$" immersion $(f,h)$ we introduce the function $\tau_h : C \longrightarrow \mathbb{Z}/4$ which measures the twisting of the image of a tubular neighborhood of the element $c \in C$.

For any $c \in C$ let $(\phi_c,\psi_c,g_c)$ be the $C^2$PN of $f$ along $c$ (see §2.3). For any $(f,h) \in I$ define $k_c = h\circ\phi_c$. In the preceding section we defined a subset $I_{V_*} \subseteq I_E$ by specifying the germs of $k_{v^*} = h\circ\phi_{v^*}$ for $v^* \in q^{-1}(v)$ and the germs $k_{c,\pm} = h\circ\phi_{c,\pm}$ at $\phi^{-1}_{c,\pm}(v^c)$ for $c$ closed and $(f,h) \in I_E$. In this section we restrict ourselves to $(f,h) \in I_{V_*}$. Since these $k_.$ map germs are fixed on $\phi_v^{-1}(q^{-1}(v) \cap S(f))$ for all vertices $v \in V$ as well as $\phi_c^{-1}(v^c)$ for $c$ closed, using the maps $\phi_v^{-1}\circ\phi_c$ or $\phi^{-1}_{c,-}\circ\phi_{c,+}$ we find that membership in $I_{V_*}$ fixes the maps $k_c$ on neighborhoods of the ends of $I \times \{0\} \subseteq I \times T(c)$. By our definition of $I_{V_*}$, the germ of $k_c : I \times T(c) \longrightarrow \mathbb{R}^2$ along $I \times \{0\}$ is fixed at $(-1,-1+\alpha)$ and $(1-\alpha,1) \times \{0\}$. Along each of these intervals the diffeomorphism germ $k_c \mid u \times T(c)$ at $(u,0)$ is constant, independent of $u$.

<u>Definition</u>. Let $k : [a,b] \times \mathbb{D} \longrightarrow \mathbb{R}^2$ be a smooth map such that on $[a,b] \times \{0\}$, $\tilde{J}(k)(u)$, the Jacobian of $k \mid \{u\} \times \mathbb{R}^2$ at $(u,0)$ is non-singular. Let $\theta(k) : [a,b] \longrightarrow \mathbb{R}$ be the smooth map defined

by $\theta(k)(a) = 0$, $Jk(u)\cdot Jk(a)^{-1} = e^{2\pi i\theta(k)(u)}$ where $\tilde{J}k : [a,b] \longrightarrow O(2)$ is the orthogonal factor of $Jk$ in its polar decomposition and we identify $SO(2)$ with $\mathbb{S}^1 \subseteq \mathbb{C}$. We define $\tau(k) = \theta(k)(b)$.

Definition. For $(f,h) \in I_{V_*}$ and $c \in C$ define

$$\tau_h(c) = \tau(k_c \mid [-1+\alpha',1-\alpha'] \times \mathbb{D}),$$

where $\alpha' \in (0,\alpha)$.

Remark. $\tau_h$ is also defined for $(f,h) \in I_V$ since for closed $c$, the integral value of $\tau_h(c)$ does not depend on fixing the germ of $h$ at any point of $c$.

By a case by case study of the expressions for $k_c$ near the end points of $I \times \{0\}$ we find that the only values that $\tau_h$ can assume are $n$, $n \pm 1/4$, $n + 1/2$ for $n \in \mathbb{Z}$.

Definition. For $(f,h) \in I_V$ define $\nu_h : C \longrightarrow \mathbb{Z}$, the integral part of $\tau_h$ by $\nu_h(c)-\tau_h(c) \in [-1/2,1/2)$ for all $c \in C$. Call $\tau_h - \nu_h$, the fractional part of $\tau_h$.

As can be seen from the canonical forms of $h$ in neighborhoods of points of $q^{-1}(V) \cap S(f)$ for $(f,h) \in I_V$ (see §3.4), the fractional part of $\tau_h$ is completely determined by the value of $\delta_h$ at the vertices. Thus there is a function $\omega : \mathcal{D} \times C \longrightarrow \{0,\pm 1/4,-1/2\}$ such that for $(f,h) \in I_V$, $\omega(\delta_h,c) = \nu_h(c) - \tau_h(c)$. (We will write $\omega_\delta(c)$ for $\omega(\delta,c)$.)

Given $(\delta,\nu) \in \mathcal{D} \times (C,\mathbb{Z})$, we define a germ $h_{\delta,\nu}$ at $S(f)$ of a map of a neighborhood of $S(f)$ into $\mathbb{R}^2$ such that $(f,h_{\delta,\nu})$ is a germ at $S(f)$ of an immersion into $\mathbb{R}^4$, and $(f,h_{\delta,\nu}) \mid S(f) = E$. Further we will want to use $h_{\delta,\nu}$ as our standard germ at $S(f)$ determined by $(\delta,\nu)$. To incorporate the $\delta$-information we fix $h_{\delta,\nu}$ near the points of $V^* = q^{-1}(V_*) \cap S(f)$, using the local forms given in §3.4 for $h$ near points of $q^{-1}(V_*) \cap S(f)$ for $(f,h) \in I_{V_*}$, where we replace $\delta_h$ by the given $\delta$ in the equations of those local forms. To complete the description of $h_{\delta,\nu}$ we define it on a disc neighborhood of each arc of $S(f)$ joining points of $V^*$ lying above an element $c \in C$.

Letting $\tau = \tau(\delta,\nu) = \nu - \omega_\delta$, we rotate a standard embedding of the disc, transverse to the arc, through an angle of $2\pi\tau(c)$ as we travel along the arc. Since the center of each disc is a point of $S(f)$, its image is determined by the requirement that $(f,h_{\delta,\nu}) \mid S(f) = E$; our rotation of that disc is about that center. Further our embedding of the transverse disc will be orientation preserving or reversing according to the sign of $\delta(c)$. In melding the map on these tubular neighborhoods of the arcs of $S(f)$ above elements of $C$ with the given maps in the neighborhoods of points of $V^*$ it is clearly no restriction to assume that the maps are undisturbed in the neighborhoods of points of $V^*$.

Since for any germ $h$ at $S(f)$ such that $(f,h)$ is the germ of an immersion in $\mathbb{R}^4$, we have functions $\nu_h$ and $\delta_h$, the equations $\nu = \nu_{h_{\delta,\nu}}$ and $\delta = \delta_{h_{\delta,\nu}}$ make sense and are valid.

__Definition__. Let $H_S$ be the set of all germs $h_{\delta,\nu}$ at $S(f)$, for $(\delta,\nu) \in \mathcal{D} \times (C,\mathbb{Z})$.

The set $I_S$ is the set of all $(f,h) \in I_{V_*}$ such that the germ of $h$ at $S(f)$ is in $H_S$.

Our object in the rest of this section is to show:

(i) If, $(f,h_0)$ and $(f,h_1) \in I_S$ and are f-regularly homotopic $(f,h_t) \in I_V$, then there is an f-regular homotopy $(f,h_t')$ joining $(f,h_0)$ to $(f,h_1)$ with $(f,h_t') \in I_S$.

(ii) If $(f,h) \in I_V$, then there is an f-regular homotopy $(f,h_t) \in I_V$ with $h_0 = h$ and $(f,h_1) \in I_S$.

To facilitate the writing we introduce two bits of notation: __If__ $Z$ __is any set interior to the domain of a function__, $g$, __we let__ $(g)_Z$ __be the germ of__ $g$ __at__ $Z$.
__If__ $I_A \subseteq I_B \subseteq I$, __we denote by__ $[I_A]_B$ __the set of all__ f-__regular homotopy classes of elements of__ $I_A$ __where the homotopy joining equivalent elements lies completely in__ $I_B$.

If $I_B$ is defined by restricting the germ of $h$ on a set $\hat{B} \subseteq M$, then we require, in addition, that $(f,h)$ and $(f,h') \in I_A$ define the same element in $[I_A]_B$ if there is a f-fibre regular homotopy $H$ joining $h$ and $h'$ such that the germ equation $(H)_{\hat{B}\times I} = (h\times 1)_{\hat{B}\times I}$ holds.

If $(f,h) \in I_A$ we denote by $[f,h]_B$ its class in $[I_A]_B$. These conventions include the special case of $[I_A]$, the f-regular homotopy classes of elements in $I_A$ where the homotopy joining equivalent elements lies in $I$.

In this language our stated objectives are to prove the injectivity of

$$0 \dashrightarrow [I_S]_S \longrightarrow [I_S]_S$$

and the surjectivity of

$$[I_S]_V \longrightarrow [I_V]_V \dashrightarrow 0$$

The following is trivial:

*Proposition 0.* *If* $I_A \subseteq I_B \subseteq I_C \subseteq I$ *then the following diagram commutes and is exact.*

$$\begin{array}{ccccc}
0 & \longrightarrow & [I_A]_B & \xrightarrow{i_B} & [I_B]_B \\
 & & \Big\downarrow{\scriptstyle j_A} & & \Big\downarrow{\scriptstyle j_B} \\
0 & \longrightarrow & [I_A]_C & \xrightarrow{i_C} & [I_B]_C \\
 & & \Big\downarrow & & \Big\downarrow \\
 & & 0 & & 0
\end{array}$$

*where* $i_B$ *and* $i_C$ *are the obvious inclusions coming from the inclusion of* $I_A \subseteq I_B$ *and* $j_A$ *and* $j_B$ *as the obvious projections coming from the inclusion* $I_B \subseteq I_C$. *If* $i_B$ *is surjective so is* $i_C$ *and if* $j_B$ *is injective so is* $j_A$.

*Proposition 1.* (i) $[I_S]_S \longrightarrow [I_V]_V$ *is bijective.*

(ii) $[I_S]_{S_*} \longrightarrow [I_S]_{V_*}$ *is bijective.*

*Corollary.* $[I_S]_E \xrightarrow{i_E} [I_{V_*}]_E$ *and* $[I_S] \xrightarrow{i} [I_{V_*}]$ *are bijective.*

*Proof.* The corollary follows from the proposition since the map $[I_S]_S \longrightarrow [I_V]_V$ factors as $[I_S]_S \xrightarrow{j_S} [I_S]_V \xrightarrow{i_V} [I_V]_V$. Since $j_S$ is surjective and $i_V$ is injective if $i_V \circ j_S$ is bijective so are $i_V$ and $j_S$. The surjectivity of $i_E$ and $i$ follow from that of $i_V$ using Proposition 0.

*Proof of Proposition 1.* Notice that $[I_S]_S \longrightarrow [I_S]_{V_*}$ and $[I_S]_{V_*} \longrightarrow [I_S]_V$ are both surjective. If we knew that $[I_S]_S \longrightarrow [I_S]_V$ were injective we would have conclusion (ii). If in addition we knew $[I_S]_V \longrightarrow [I_V]_V$ were surjective we would have (i) as well. Thus we break the proof into two parts:

A) $i_V : [I_S]_V \longrightarrow [I_V]_V$ is surjective.

B) $j_S : [I_S]_S \longrightarrow [I_S]_V$ is injective.

To prove A) it suffices to show if $(f,h) \in I_V$ we can find $(f,h_1)$ $I_S$ such that $[f,h_1]_V = [f,h]_V$.

To prove B) it suffices to show given $(f,h_0)$ and $(f,h_1) \in I_S$ for which there is a homotopy $H : h_0 \simeq h_1$ with $(f,h_t) \in I_V$ then there is a homotopy $H' : h_0 \simeq h_1$ such that $(H')_{S(f)\times I} = (h_0\times 1)_{S(f)\times I}$.

For both of these statements it will suffice to work with the set of mappings, $K_J \subseteq C^\infty(I\times J\times \mathbb{D}, \mathbb{R}^2)$ where $I = [0,1]$ and $J$ is a connected manifold possibly with boundary, where $k \in K_J$ satisfies:

1) $1 \times k : I \times J \times \mathbb{D} \longrightarrow I \times J \times \mathbb{R}^2$ is an embedding

2) $k \mid I \times J \times \{0\} = 0$

3) There is a neighborhood $U$ of $\partial I$ such that $k(u,v,(x,y))$ is independent of $v \in J$ for all $u \in U$.

In case $J = \{*\}$ a single point we identify $I \times \{*\} \times \mathbb{D}$ with $I \times \mathbb{D}$ and call $K_*$ simply $K$.

For any $k \in K_J$ we define for each $v \in J$, $\tau(k_v)$ as usual. By property 3) of $k \in K_J$, $\tau(k_v) - \tau(k_{v'})$ is integral for $v, v' \in J$, thus $\tau(k_v)$ is constant since $J$ is connected. We denote this common value of $\tau(k_v)$ simply as $\tau(k)$.

To prove A) and B) we prove:

Lemma.

a) Let $k_0, k_* \in K_J$ and suppose that $k_0$ and $k_*$ agree on $(U\times J\times \mathbb{D}) \cup (I\times V\times \mathbb{D})$ where $U$ is a neighborhood of $\partial I$ and $V$ is a neighborhood of $\partial J$. Suppose further that $\tau(k_0) = \tau(k_*)$, then there is a homotopy, $K$, fixed in a neighborhood of $\partial(I\times J\times \mathbb{D})$ with $K_t \in K_J$ for all $t \in [0,1]$ and $k_0 = K_0$ and on a neighborhood of $I \times J \times \{0\}$, $K_1 = k_*$.

b) Let $k \in K$ and suppose there is a homotopy $K^0 : k \simeq k$ which is fixed on a neighborhood $U \times \mathbb{D}$ of $\partial I \times \mathbb{D}$ and such that $K_t^0 \in K$ for all $t \in [0,1]$. Then there is a homotopy $K : k \simeq k$ such that for all $t \in [0,1]$, $K_t = K_t^0$ in a neighborhood of $\partial(I\times \mathbb{D})$ and which is constantly $k$ in a neighborhood of $I \times \{0\}$.

That Lemma a) implies A) is easy since if $(f,h) \in I_V$, Lemma a) allows us to construct a regular homotopy $(f,h_t) \in I_V$ where $h_t$ is constant on a neighborhood of the boundary of a neighborhood of the arcs of $S(f)$ such that $h_1$ is canonical in a neighborhood of the arcs of $S(f)$, i.e. $(f,h_1) \in I_S$. Here we use the version of a) in which $k_0, k_* \in K$.

To see that Lemma b) implies B) we need:

Proposition 2. $\tau : I_V \longrightarrow T$ is constant on classes in $[I_V]_V$.

Proof. It is obviously enough to prove:

Lemma. Let $k_i \in K_J$, $i = 0, 1$ with $k_0 = k_1$ on a neighborhood $U \times J \times \mathbb{D}$ of $\partial I \times J \times \mathbb{D}$. If $K : k_0 \simeq k_1$ with $K \in K_{J\times T}$ where $T = [0,1]$ then $\tau(k_0) = \tau(k_1)$.

Proof of Lemma. By the remarks following the introduction of $K_J$, we know that $\tau(K)$ is the common value of $\tau(K_{v,t})$ for all $(v,t) \in J \times T$, and $\tau(K_t)$ is the common value of $\tau(K_{v,t})$ for all $v \in J$. But $K_0 = k_0$, and $K_1 = k_1$. Hence $\tau(k_0) = \tau(k_1) = \tau(K)$. //

Thus suppose $(f,h_0)$ and $(f,h_1) \in I_S$ and $[f,h_0]_V = [f,h_1]_V$ as in B). Since $[f,h_0] = [f,h_1]$, we know that $\delta_{h_0} = \delta_{h_1}$ and since $[f,h_0]_V = [f,h_1]_V$ we know by Proposition 2 that $\tau_{h_0} = \tau_{h_1}$. Now since $(f,h_0)$ and $(f,h_1)$ are in $I_S$ we know that the germs of $h_0$ and $h_1$ coincide along $S(f)$. Lemma b) allows us to replace the homotopy in $I_V$ between $(f,h_0)$ and $(f,h_1)$ with a homotopy in $I_S$. The new homotopy differs from the given one only in a neighborhood of the arcs of $S(f)$ outside of neighborhoods of the points of $q^{-1}(V) \cap S(f)$.

Proof of Lemma a). For $k \in K_J$, we let $\tilde{J}(k)(u,v)$ be the Jacobian of $k \mid ((u,v)) \times \mathbb{D})$ at $(u,v,0)$. Define $\ell : I \times J \times \mathbb{D} \longrightarrow \mathbb{R}^2$ by $1 \times \ell = (1\times k_*)^{-1} \circ (1\times k_0)$. Since $\ell \mid I \times J \times 0 = 0$ and $\ell \mid (U\times J \cup I\times V) \times \mathbb{D}$ is projection on $\mathbb{D}$, $\ell \in K_J$. Notice that $\tilde{J}(\ell) : I \times J \longrightarrow GL(2,\mathbb{R})$ is the identity in a neighborhood of $\partial(I\times J)$. Thus there is a smooth homotopy, constantly the identity on a neighborhood of $\partial(I\times J)$, $L : J(\ell) \simeq id_{\mathbb{R}^2}$. Define $R : I \times J \times \mathbb{D} \longrightarrow \mathbb{R}^2$ by $R(u,v,x) = \ell(u,v,x) - \tilde{J}(\ell)(u,v)x$, and define the homotopy $H : T \times I \times J \times \mathbb{D} \longrightarrow \mathbb{R}^2$ by $H(t,u,v,x) = L(t,u,v)x + (1-t)R(u,v,x)$ for $T = [0,1]$. Note that $H_0 = \ell$, $H_1 = proj_{\mathbb{R}^2}$, projection into $\mathbb{R}^2$, $H \mid T \times I \times J \times 0 = 0$. Since $\ell \mid (U\times J \cup I\times V) \times \mathbb{D} = proj_{\mathbb{R}^2}$, we know that $R \mid (U\times J \cup I\times V) \times \mathbb{D} = 0$. Hence $H \mid T \times (U\times J \cup I\times V) \times \mathbb{D} = proj_{\mathbb{R}^2}$. We apply Lemma 0 of §2.1 to $H$ obtain concentric discs about $0$, $\mathbb{D}_0 \subseteq \mathbb{D}_1 \subseteq \mathbb{D}$ and neighborhoods $U_1$ and $V_1$ such that $\partial I \subseteq U_1 \subseteq U$ and $\partial J \subseteq V_1 \subseteq V$ and a diffeomorphism

$$1 \times \tilde{H} : (T\times I\times J) \times \mathbb{R}^2 \longrightarrow (T\times I\times J) \times \mathbb{R}^2$$

such that

1) On $T \times I \times J \times \mathbb{D}_0$, $\tilde{H} = H$
2) On $T \times (I\times J\times\mathbb{R}^2 - (I-U_1)\times(J-V_1)\times\mathbb{D}_1)$, $\tilde{H} = proj_{\mathbb{R}^2}$.

Claim: Our desired homotopy is given by $1 \times K = (1\times k_0)\circ(1\times\tilde{H}_0)^{-1}\circ(1\times\tilde{H})$. Obviously $1 \times K_0 = 1 \times k_0$. For each $t \in T$, $K_t \in K_J$ since $k_0$ and $\tilde{H}_t \in K_J$. From 2) above, $1 \times \tilde{H}_t$ = identity on a neighborhood of $\partial(I\times J\times \mathbb{D})$, so on that neighborhood $1 \times K_t = 1 \times k_0$ for all $t \in T$. Finally by 1) above on a neighborhood of $I \times J \times \{0\}$, $\tilde{H}_0 = H_0 = \ell$ and $\tilde{H}_1 = H_1 = \text{proj}_{\mathbb{R}^2}$. Thus on that neighborhood of $I \times J \times \{0\}$, $1 \times K_1 = (1\times k_0)\circ(1\times\ell)^{-1} = 1 \times k_*$. // Lemma a)

Proof of Lemma b). Let $T = [0,1]$. The homotopy $K^0 : I \times T \times \mathbb{D} \longrightarrow \mathbb{R}^2$ is an element of $K_T$. Let $K^* : I \times T \times \mathbb{D} \longrightarrow \mathbb{R}^2$ be given by $K^*(u,t,x) = k(u,x)$; $K^* \in K_T$. Obviously $\tau(K^0) = \tau(K^*) = \tau(k)$. We apply Lemma a) to $K^0$ and $K^*$ and obtain a homotopy $\mathbb{K}$, fixed on a neighborhood of $\partial(I\times T\times\mathbb{D})$ with $\mathbb{K}_s \in K_T$ for all $s \in [0,1]$, $\mathbb{K}_0 = K^0$ and on a neighborhood of $I \times T \times \{0\}$, $\mathbb{K}_1 = K^*$.

We check that $\mathbb{K}_1$ is the desired homotopy. Since $\mathbb{K}$ is fixed in a neighborhood of $\partial(I\times T\times\mathbb{D})$, $\mathbb{K} \mid I \times j \times \mathbb{D} = \mathbb{K}_0 \mid I \times j \times \mathbb{D} = K^0 \mid I \times j \times \mathbb{D} = k \mid I \times \mathbb{D}$ for $j = 0, 1$; thus $\mathbb{K}_1 : k \simeq k$. Further for all $t$, on a neighborhood of $\partial(I\times\mathbb{D})$, $(\mathbb{K}_1)_t = K^0_t$. Finally for each $t \in T$, in a neighborhood of $I \times \{0\}$, $(\mathbb{K}_1)_t = K^*_t = k$.

// Lemma b)

If we combine the Corollary to Proposition 1 with the Proposition and Corollary of §3.4 we conclude that:

$$[I_S] = [I_V] = [I_E] = [I] \quad\text{and}\quad [I_S]_E = [I_V]_E = [I_E]_E.$$

As we already remarked, $\tau$ is not constant on classes of $[I_V]_E$; however, by Proposition 2, $\tau$ is well defined on $[I_V]_V$. Thus we know that $j_V : [I_V]_V \longrightarrow [I_V]_E$ is not injective. We will study this lack of injectivity in §3.10.

## 3.6 Standard forms for germs of lifts at $\hat{\Sigma}$, $[I_\Sigma]_\Sigma = [I_S]_S$.

In the previous paragraph we showed that $[I_V]_V$ and $[I_S]_S$ are in 1:1 correspondence, so in studying $[I_V]_V$ there is no loss of generality to restrict our attention to those $(f,h) \in I_V$ such that the germ of $h$ at $S(f)$, $(h)_{S(f)} \in H_S$.

The germs of $H_S$ are in 1:1 correspondence with the element of $\mathcal{D} \times (C,\mathbb{Z})$.

In this paragraph we will show that we can further restrict the maps $(f,h) \in I_S$ by making the germ of $h$ on $\hat{\Sigma}$ standard. (Recall that $\hat{\Sigma} = q^{-1}q(S(f))$.

We give an indication of what $\hat{\Sigma}$ looks like in the neighborhood of the q-fibre of a vertex.

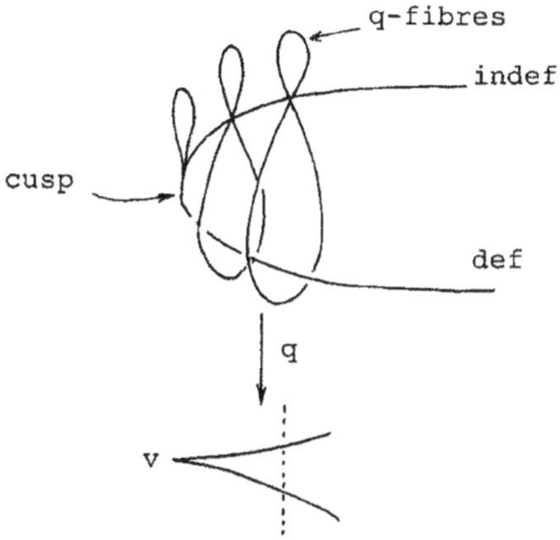


v, a cusp

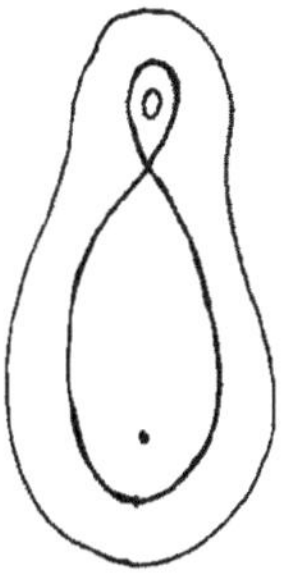

transverse manifold
above the dotted line

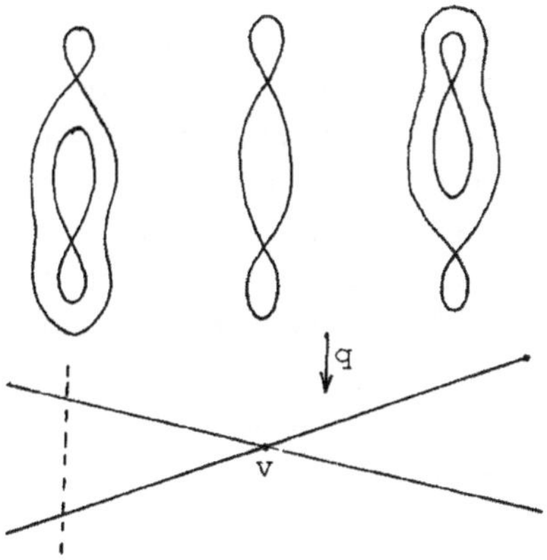


v, a (1·2·2·3) point

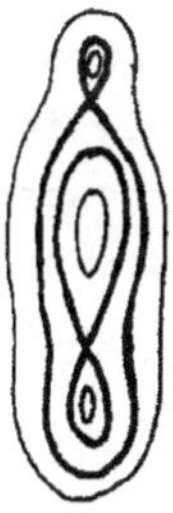

transverse manifold
above the dotted line

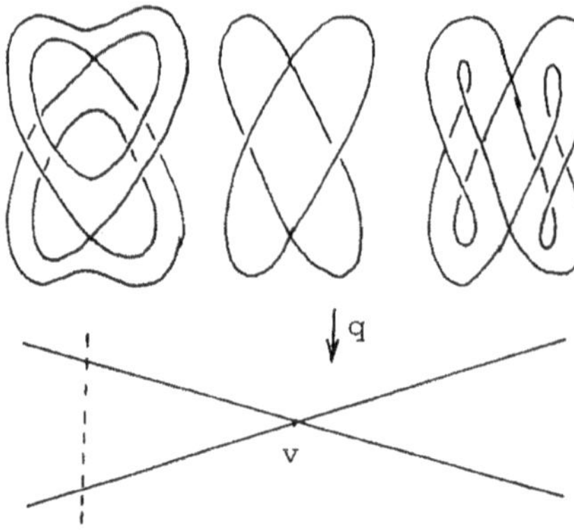


v, a (1·2·2·1) point

transverse manifold above the dotted line

Notation. If for a subset $\hat{A} \subseteq M$, $H_A$ is a set of germs of maps of a neighborhood of $\hat{A}$ into $\mathbb{R}^2$, and if $\hat{A} \subseteq \hat{B}$, then we denote by $H_{A,B}$ the set of those germs at $\hat{B}$ of maps into $\mathbb{R}^2$ whose germs at $\hat{A}$ are in $H_A$.

Having defined $H_A$ for some $\hat{A} \subseteq M$, we define $I_A \subseteq I_E$ by $(f,h) \in I_A$ iff $(h)_{\hat{A}} \in H_A$. If we let $\hat{V}_* = V^* = q^{-1}(V_*) \cap S(f)$, $\hat{V} = q^{-1}(V) \cap S(f)$, and $\hat{S} = S(f)$ it is clear from the definitions of $I_{V_*}$, $I_V$ and $I_S$ in §3.3 and §3.4 how to define $H_{V_*}$, $H_V$ and $H_S$ so as to conform to the notation described above.

In this paragraph we define $H_\Sigma \subseteq H_{S,\Sigma}$ and will show that $[I_\Sigma]_\Sigma$ and $[I_S]_S$ are in 1:1 correspondence. We will also show that $H_\Sigma$ is in 1:1 correspondence with a subset of $D \times (C \cup A, \mathbb{Z})$, where $A$ is the set of arcs of $q^{-1}(\xi) - S(f)$ for $\xi \in \Sigma = q(S(f))$.

We recall the definition of the rotation number of any immersion $\mu : [a,b] \longrightarrow \mathbb{R}^2 = \mathbb{C}$. Let $\theta : [a,b] \longrightarrow \mathbb{R}$ be the function uniquely defined by $\theta(a) \in [0,1)$ and $\mu' = |\mu'|e^{2\pi i\theta}$. The rotation number of $\mu$, $\operatorname{rot} \mu = \theta(b) - \theta(a)$. Obviously $\operatorname{rot} \mu = \operatorname{rot} \mu_1$ if $\mu_1$ is an orientation-preserving reparametrization of $\mu$. We write $n(\mu) = \operatorname{rot} \mu + w(\mu)$ where $n(\mu) \in \mathbb{Z}$ and $w(\mu) \in [-1/2,1/2)$.

For any $\xi \in \Sigma$, let $A_\xi$ be the collection of oriented component arcs of $q^{-1}(\xi) - S(f)$. We picture $q^{-1}(\xi)$ as it appears in the transverse manifold at $q^{-1}(\xi)$:

$\xi \in C_0$ $\xi \in C_1$ $\xi$, a cusp $\xi$, $(1\cdot 2\cdot 2\cdot 1)$ $\xi$, $(1\cdot 2\cdot 2\cdot 3)$

The heavy points are those of $q^{-1}(\xi) \cap S(f)$. Obviously any arc of $A_\xi$ can be parametrized by an immersion $\alpha : (0,1) \longrightarrow q^{-1}(\xi)$ which extends to an immersion $\bar{\alpha} : [0,1] \longrightarrow q^{-1}(\xi) \subseteq M$, where $\bar{\alpha}(0)$ and $\bar{\alpha}(1)$ are in $q^{-1}(\xi) \cap S(f)$. We will, by an abuse of language, refer to $\alpha$ as an element of $A_\xi$. Let $A = \bigcup_{\xi \in \Sigma} A_\xi$.

Definition. For $(h) \in H_{S,\Sigma}$, $\sigma_h : A \longrightarrow \mathbb{R}$ is defined by $\sigma_h(\alpha) = \mathrm{rot}(h \circ \bar{\alpha})$ for $\alpha \in A$.

In §3.5 we defined a map $\nu : I_V \longrightarrow (C, \mathbb{Z})$ and $\omega : \mathcal{D} \times C \longrightarrow [-1/2, 1/2)$ such that $\tau_h = \nu_h - \omega_{\delta_h}$. Both $\tau$ and $\nu$ depend only on the germ $(h)_{S(f)}$ for $(f,h) \in I_V$. Thus restricting to $I_S$ we may consider $\nu : H_{S,\Sigma} \longrightarrow (C, \mathbb{Z})$, since $H_S$ is the collection of all germs $(h)_{S(f)}$ for $(f,h) \in I_S$. We now define

$\nu : H_{S,\Sigma} \longrightarrow (C \cup A, \mathbb{Z})$ and $\omega : \mathcal{D} \times (C \cup A) \longrightarrow [-1/2, 1/2]$ as extensions of the functions already defined as follows: For any $(h) \in I_{S,\Sigma}$ and $\alpha \in A_\xi$, write $\sigma_h(\alpha) + \omega_h(\alpha) = \nu_h(\alpha)$, where $\nu_h(\alpha) = n(h \circ \bar{\alpha})$, and $\omega_h(\alpha) = w(h \circ \bar{\alpha})$ except if $\xi$ is a cusp. In that case $A_\xi$ has only one element and we define $\omega_h(\alpha) = (1/2)\delta_h(\xi)$. This gives the extension of $\nu$. To obtain the extension of $\omega$, we must show that the $\omega_h : A \longrightarrow [-1/2, 1/2]$ defined above depends only on $\delta_h$.

Proposition 0. For $(h)$ and $(h')$ in $H_{S,\Sigma}$, $\omega_h = \omega_{h'}$ iff $\delta_h|C_1 = \delta_{h'}|C_1$.

Proof. Suppose for some $c \in C_1$, that $\delta_h(c) = -\delta_{h'}(c)$. Then for every arc $\alpha$ in $q^{-1}(\xi)$ for $\xi \in c$, $\omega_h(\alpha) = -\omega_{h'}(\alpha)$ since the h-images of the initial and terminal vectors of $\bar{\alpha}$ are mirror images of their h'-images.

Conversely, if $\delta_h|C_1 = \delta_{h'}|C_1$, we know that the maps h and h' in a neighborhood of $S(f)$ are completely determined by their $\delta$'s and $\tau$'s. In a neighborhood of $q^{-1}(V) \cap S(f)$, h and h' are completely determined by the $\delta$'s alone. Hence for arcs of $A$ which end at points of $S(f)$ near the $q$-preimage of a vertex, $\omega_h$ and $\omega_{h'}$ agree. For $c \in C_1$, let $(\phi_c, \psi_c, g_c)$ be the $C^2$PN of $c$ (see §2.3), and let $k = h \circ \phi_c$, $k' = h' \circ \phi_c$. The maps, $k$ and $k'$, take $I \times T(c)$ into $\mathbb{R}^2$. The values of $\omega_h(\alpha)$ and $\omega_{h'}(\alpha)$ for any $\alpha \in A_\xi$ for $\xi \in c$ are determined by the maps $k$ and $k'$ on a neighborhood of $(u,0)$ in $u \times T(c)$ where $\phi_c(u,0) = \xi$. However the maps $k$ and $k'$ are completely determined by $\delta_h$ and $\delta_{h'}$ on $(I_\pm \times \mathbb{D}) \subseteq I \times T(c)$, for $\mathbb{D}$ some disc about $0$. Further for all $u$ in $I - (I_+ \cup I_-)$, $k$ and $k'$ on $u \times \mathbb{D}$ is just the composition of a map that depends only on $k$ and $k'$ on $I_- \times \mathbb{D}$, followed by a rotation on a little transverse disc at $(u,0)$ (see §3.5 for the definition of $I_S$). So the angle between the initial and terminal vectors of the $h_-$ and $h'_-$ images of $\alpha$ depend only on $\delta_h(c)$ and $\delta_{h'}(c)$. //

Remark: Let $c \in C_1$ and let $\xi \in c$. For any sufficiently small connected neighborhood $U$ of $\xi$ in $c$, if we choose a component of $q^{-1}(U) - S(f)$, there will be a unique arc $\alpha_{\xi'}$ in that component for all $\xi' \in U$. Except in a neighborhood of a cusp, $\omega_h(\alpha_{\xi'})$ will be independent of $\xi' \in U$. If $c$ abuts a cusp, as $\xi$ approaches the cusp along $c$, the $\omega_h$-values increase from $1/4$ to $1/2$ if $\delta_h(c) = 1$ and decrease from $-1/4$ to $-1/2$ if $\delta_h = -1$. We now have functions: $\omega : \mathcal{D} \times (C \cup A) \longrightarrow [-1/2, 1/2]$ and $\nu : H_{S,\Sigma} \times (C \cup A) \longrightarrow \mathbb{Z}$ such that for $h \in H_{S,\Sigma}$

$$\tau_h = (\nu_h - \omega_{\delta_h}) \mid C \quad \text{and} \quad \sigma_h = (\nu_h - \omega_{\delta_h}) \mid A.$$

Just as we defined $\tau(\delta,\nu) = (\nu - \omega_\delta) \mid C$ so we define $\sigma(\delta,\nu) = (\nu - \omega_\delta) \mid A$ for any $\delta \in \mathcal{D}$ and any $\nu \in (C \cup A, \mathbb{Z})$.

Definition. Let $\lambda : H_{S,\Sigma} \longrightarrow \mathcal{D} \times (C \cup A, \mathbb{Z}) : (h) \longrightarrow (\delta_h, \nu_h) = \lambda_h$ and let $L_\Sigma = \lambda(H_{S,\Sigma})$.

Let $H_\Sigma$ be any subset of $H_{S,\Sigma}$ such that $\lambda$ is a 1:1 correspondence between $H_\Sigma$ and $L_\Sigma$. This $H_\Sigma$ is our set of model map-germs at $\hat{\Sigma}$ and $L_\Sigma$ indexes them. In the last section of this chapter we will choose $H_\Sigma$ a little less arbitrarily. As usual, we let

$$I_\Sigma = \{(f,h) \in I_S \mid (h)_{\hat{\Sigma}} \in H_\Sigma\}.$$

The principal object of this section is to prove:

Proposition 1. The (obvious) map $[I_\Sigma]_\Sigma \longrightarrow [I_S]_S$ is a 1:1 correspondence.

We prove: a) the injection $[I_\Sigma]_S \longrightarrow [I_S]_S$ is a bijection and b) the projection $[I_\Sigma]_\Sigma \longrightarrow [I_\Sigma]_S$ is a bijection.

To prove that $[I_\Sigma]_S \longrightarrow [I_S]_S$ is a bijection it suffices to prove:

Proposition 2. Let $(h_1)$, $(h_0) \in H_{S,\Sigma}$. The following are equivalent:

1) $\lambda_{h_0} = \lambda_{h_1}$.

2) If $h_0$ and $h_1$ are any representatives of $(h_0)$ and $(h_1)$ with a common domain $B(\Sigma)$, then there is a smooth homotopy $H : h_0 = H_0 \simeq H_1$ such that for all $t \in [0,1]$

(i) $(f,H_t)$ immerses $B(\Sigma)$ in $\mathbb{R}^4$.

(ii) $H_t$ is independent of $t$ in a neighborhood of $S(f)$ and on a neighborhood of $\partial B(\Sigma)$.

(iii) $(H_1)_{\hat{\Sigma}} = (h_1)_{\hat{\Sigma}}$.

The proof of the proposition is a consequence of a generalization of the Whitney-Graustein Theorem classifying immersions of circles in the plane under regular homotopy [W-G]. The details are given in the Appendix.

We use the same technique to complete the proof of Proposition 1.

Proof. Let $h : B(\Sigma) \longrightarrow \mathbb{R}^2$, $(h)_{\hat{\Sigma}} \in H_\Sigma$ and suppose $H^0 : h \simeq h$ such that

1) $(f,H^0_s)$ immerses $B(\Sigma)$ in $\mathbb{R}^4$

2) $H^0_s$ is independent of $s$ in a neighborhood of $S(f)$.

To prove that the projection $[I_\Sigma]_\Sigma \longrightarrow [I_\Sigma]_S$ is a bijection it suffices to show that we can find a homotopy $H^1 : h \simeq h$ such that

1) $(f,H^1_s)$ immerses $B(\Sigma)$ in $\mathbb{R}^4$

2) $H^1 = H^0$ on a neighborhood of $I \times \partial B(\Sigma)$

3) On a neighborhood of $\hat{\Sigma}$, $H^1_s = h$ for all $s$.

We show that we can obtain $H^1$ as the end map of a homotopy that begins with $H^0$; we find a homotopy $\mathbb{H} : H^0 \simeq H^1$, $\mathbb{H} : (I\times B(\Sigma)) \times T \longrightarrow \mathbb{R}^2$, $T = [0,1]$ such that

(i) $(f,\mathbb{H}_t) \mid s \times B(\Sigma)$ is an immersion in $\mathbb{R}^4$ for all $s, t \in I \times T$.

(ii) $\mathbb{H}_t$ is independent of $t$ in a neighborhood of $I \times S(f)$ and in a neighborhood of $I \times \partial B(\Sigma)$.

(iii) $(\mathbb{H}_1)_{I\times\hat{\Sigma}} = (H^1)_{I\times\hat{\Sigma}}$.

The proof of the existence of $\mathbb{H}$ is analogous to that of statement 2) of Proposition 2 with an extra $I = [0,1]$-factor in the domain of all mappings. // (Proposition 1)

By an abuse of language let, $\lambda : I_S \longrightarrow L_\Sigma : (f,h) \longrightarrow \lambda(h)_{\hat{\Sigma}}$. Proposition 2 shows that $\lambda$ is constant on classes, $[I_S]_S$ and also:

Corollary 1. If $(f,h_0), (f,h_1) \in I_S$ then $\lambda(f,h_0) = \lambda(f,h_1)$ iff there is an element $(f,h) \in [f,h_0]_S$ such that the germs $(h)_{\hat{\Sigma}} = (h_1)_{\hat{\Sigma}}$.

Thus the value of $\lambda(f,h)$ determines the f-regular homotopy class, keeping the germ fixed at $S(f)$, of the germ $(f,h)_{\hat{\Sigma}}$.

## 3.7 Relations among $\{r,\delta,\sigma\}$--those that arise from lifting $f \mid B(\Sigma)$

In §3.1 we introduced a set $\mathrm{Rot} \subseteq (R,\mathbb{Z})$ and a map from $[I]$ into Rot which takes $[f,h]$ into $r_h$. For each $R \in R$, $r_h(R)$ is the rotation number of $h$ on the circle fibre of $B(R)$, and is part of the characterization of the f-fibre regular homotopy class of $h \mid B(R)$. In §3.2, it was shown that for $(f,h) \in I$, $r_h \in \mathrm{Rot}\,\vartheta_h$, where $\vartheta_h$ is an orientation of $C$ determined by $r_h$. Furthermore in the Lemma of §3.2 it was shown that for any $c \in C_i$, $\vartheta_h(c)\delta_h(c) = (-1)^i$, where $\delta_h \in D$ ($\delta_h(c) = +1$ $(-1)$ if $h$ maps an oriented disc, transverse to $c$ in an orientation preserving (reversing) way. Thus any $\delta \in D$ determines an orientation $\vartheta_\delta$ by $\vartheta_\delta(c)\delta(c) = (-1)^i$ if $c \in C_i$. In particular $\vartheta_h = \vartheta_{\delta_h}$. Since $\vartheta_h$ is determined by $r_h$ so also is $\delta_h$. Thus the inclusion of both $\delta_h$ and $r_h$ as invariants of f-regular homotopy is redundant. In fact we will show that $r_h$ is completely determined by $\delta_h$ and $\nu_h \mid A^*$ where $\nu_h \in (C \cup A,\mathbb{Z})$ is the map introduced in §3.5 and §3.6 and $A^* = \cup\{A_v \mid v \in V_*\}$. If $R \in R$ has an element $c \in C_0$ on its boundary, then $r_h(R) = \delta_h(c)$.

For each $R \in R$, let $v(R) = V_* \cap \bar{R}$ and for each $c \in C$, let $v(c) = V_* \cap \bar{c}$. In this section we show that for $(f,h) \in I_S$ there is a recipe $(R_1)$ for computing $\nu_h \mid A_\xi$ in terms of $\delta_h$ and $\nu_h \mid A_v$ where $\xi \in c$ and $v \in v(c)$, and a recipe $(R_2)$ for computing $r_h(R)$ in terms of $\delta_h$ and $\nu_h \mid A_v$ for any $v \in v(R)$. For any $\delta \in D$, let $N_\delta \subseteq (C \cup A^*,\mathbb{Z})$ be the set of those maps $\nu$ which extend uniquely in $(C \cup A,\mathbb{Z})$ according to $(R_1)$. Since $v(c)$ may have two elements for some $c \in C$, $\nu \in N_\delta$ satisfies a generally non-trivial condition--the First Consistency Condition.

In the previous section we let $L_\Sigma \subseteq \mathcal{D} \times (C \cup A, \mathbb{Z})$ be the image of the map $\lambda : H_{S,\Sigma} \longrightarrow \mathcal{D} \times (C \cup A, \mathbb{Z})$, where $\lambda(h) = (\delta_h, \nu_h)$. We prove:

<u>Proposition 1</u>. $(\delta,\nu) \in L_\Sigma$ <u>iff</u> $\nu \mid C \cup A^* \in N_\delta$.

Since the extension of $\nu \in N_\delta$ to an element of $(C \cup A, \mathbb{Z})$ is unique (as specified by $R_1$) it is a harmless abuse of notation if we write $L_\Sigma = \cup\{\delta \times N_\delta \mid \delta \in \mathcal{D}\}$, and consider, if convenient, $N_\delta \subseteq (C \cup A, \mathbb{Z})$.

Similarly for $\delta \in \mathcal{D}$, let $N_\delta' \subseteq N_\delta$ be the set of those maps which uniquely define maps $\tilde{r}(\delta,\nu) \in (R, \mathbb{Z})$ according to $(R_2)$. Since $\nu(R)$ may have many elements, a <u>Second Consistency Condition</u> is imposed on $\nu$ for membership in $N_\delta'$.

If we let $L_\Sigma' = \{(\delta,\nu) \in L_\Sigma \mid \nu \in N_\delta'\}$ then we know that $\lambda(I_S) \subseteq L_\Sigma'$ where $(f,h) = (\delta_h, \nu_h)$. Although this containment is generally proper we have:

<u>Proposition 2</u>. <u>For</u> $\nu \in N_\delta'$, $\tilde{r}(\delta,\nu) \in \mathrm{Rot}(\vartheta_\delta)$, <u>where</u> $\vartheta_\delta$ <u>is the orientation defined by</u> $\vartheta_\delta(c) = (-1)^i \delta(c)$ <u>for</u> $c \in C_i$, $i = 0, 1$.

In the following we will be working in the $C^2$PN's and our figures will be drawn in the transverse manifolds $T(v)$ or $T(c)$ immersed in $\mathbb{R}^2$. Given any point $\xi \in W$, we will consider the transverse manifold to any point in a neighborhood of $\xi$ to be canonically embedded in the transverse manifold of $\xi$. Suppose for example that $v$ is a vertex and $(\phi_v, \psi_v, g_v)$ is the $C^2$PN for $v$. The transverse manifold at $v$ is $\phi_v(0 \times T(v))$, which is diffeomorphic to $\phi_v(u \times T(v))$ for all $u \in I$ (see §2.2), $u \times T(v) = (1 \times g_v)^{-1}(u \times J)$. Now any point $x_0 \in \phi_v(I \times T(v))$ has f-image equal to $\psi_v(u_0, v_0)$, and the transverse arc at that point is $\psi_v(u_0 \times J_0)$ for a subinterval $J_0$ of $J$. Thus the transverse manifold at $x_0$ is $\phi_v(1 \times g_v)^{-1}(u_0 \times J_0)$ contained in $\phi_v(u_0 \times T(v))$ (see §2.3). This gives us our canonical embedding of transverse manifolds for points in a neighborhood of $\xi$ in the transverse manifold at $\xi$. Given any arc $a \in A_v$, we have a canonical arc in $\phi_v^{-1} \circ a$ in $0 \times T(v)$ which we can consider in $u \times T(v)$ for any $u \times I$. If it lies in $(1 \times g_v)^{-1}(u_0 \times J_0)$, we can consider $a$ in the transverse manifold to $x_0$, close, perhaps, to a subarc of an element of $A_{q(x_0)}$. In the sequel although all of the constructions and figures are in the $\mathbb{R}^2$ preimages of arcs in $M$ (via $\phi_v^{-1}$ or $\phi_c^{-1}$) we will generally omit the explicit diffeomorphisms from the notation.

We label the arcs (with their usual orientations) of $q^{-1}(\xi) - S(f)$ as follows:

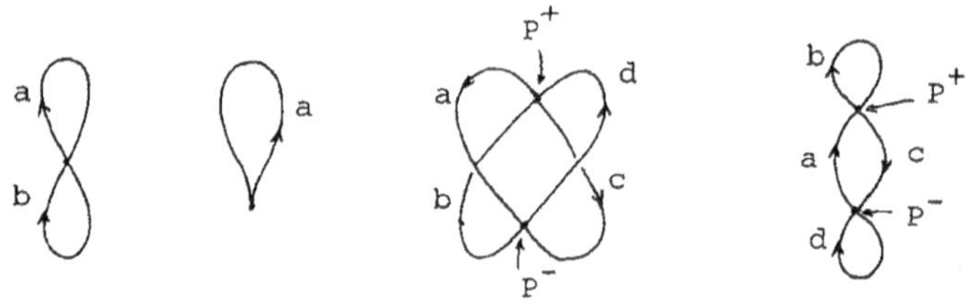


$\xi \in c \in C_1$ $\xi$, a cusp $\xi$, a (1·2·2·1) vertex $\xi$, a (1·2·2·3) vertex

The dark points are as usual points of $S(f)$ in $q^{-1}(\xi)$. This gives us a complete catalog of labels for the component arcs, the elements of $A$. We will use the notation $[a]$ instead of $\bar{a}$ to denote the extension to the closed interval of the arc, $a$ (see §3.6). We will write $\nu[a]$ instead of $\nu(a)$.

Definition. A word $a_1 \ldots a_n$ in the arc labels over a point of $\Sigma$ is called an arc-word if the terminal point of $a_i$ is the initial point of $a_{i+1}$. If, in addition, the terminal point of $a_n$ is the initial point of $a_1$ we call the word a cycle-word.

If $a_1 \ldots a_n$ is an arc-word, we denote by $[a_1, \ldots, a_n]$ the smooth oriented compact arc constructed from the sequence of $[a_i]$ obtained by traversing the arcs in order and smoothing internal transitions by replacing

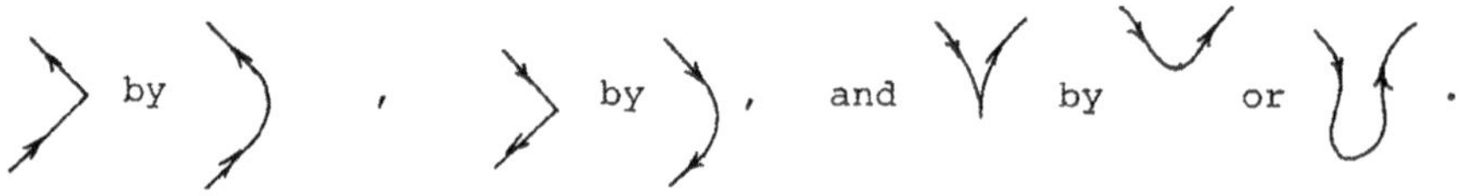


If $a_1 \ldots a_n$ is a cycle-word, we denote by $(a_1 \ldots a_n)$ the smooth closed curve obtained by smoothing the transition from $a_n$ to $a_1$ as well as all of the interval transitions.

To compare $r$ and $\nu$, we label the regions of $W - \Sigma$, near points of $\Sigma_1 = \Sigma - \cup\{c \in C_0\}$ as follows. For any point $\zeta \in W - \Sigma$ in a neighborhood of a point $\xi \in \Sigma_1$, there are closed curves in the transverse manifold to $\xi$, homotopic there, to the q-fibre over $\zeta$. These closed curves are uniquely described as $(w)$ where $w$ is a word in the arc labels for $A_\xi$. We picture neighborhoods of $\xi$ in W:

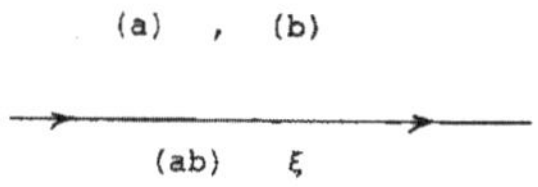


$\xi \in c \in C_1$

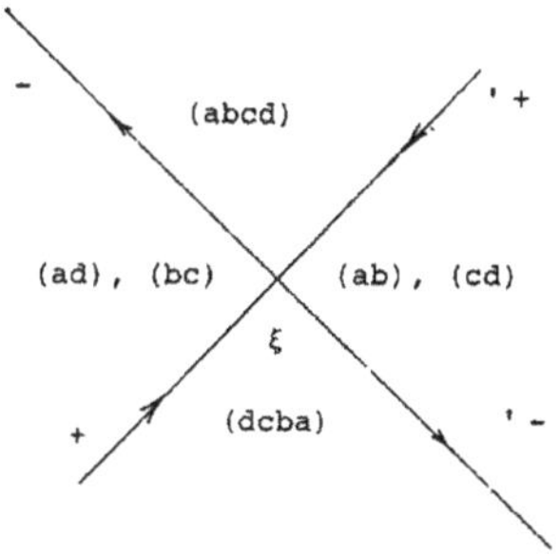


$\xi$, a (1·2·2·1) vertex

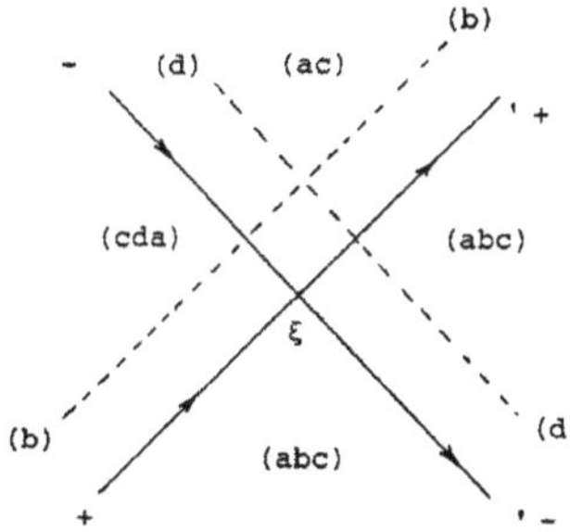


$\xi$, a (1·2·2·3) vertex

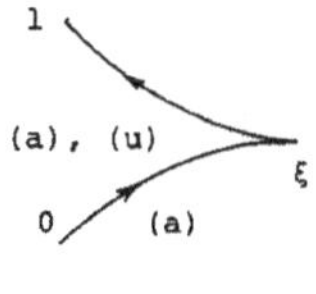


$\xi$, a (-1)-cusp ,

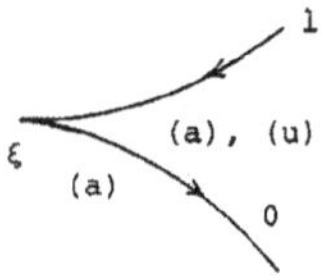


$\xi$, a (+1)-cusp

Here the many words in a single region represent the many elements of $R$ that share the same arcs of $C$ as parts of their boundaries. For example for $\xi \in c \in C_1$, the figure is simple to draw.

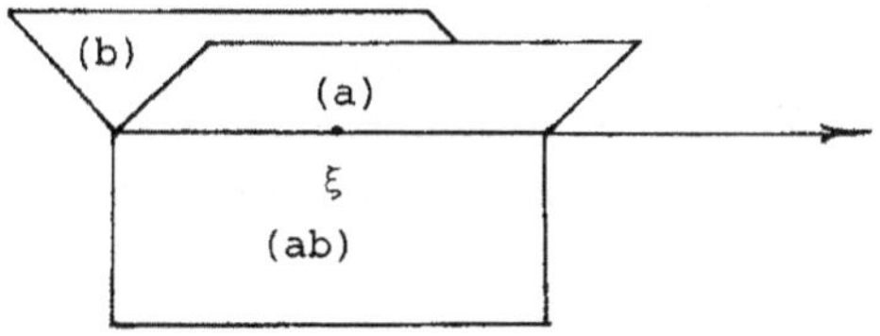


A neighborhood in W of $\xi \in c \in C_1$

In the (1·2·2·3) case the dotted line indicates that the element of $R$ whose q-fibre is homotopic to (b) or (d) covers a half space rather than a quadrant near $\xi$. In the cusp cases the (u) curves are curves that are null homotopic in the transverse manifolds to the cusps. In general, the homotopy classes of the curves are nontrivial in the transverse manifolds and generate (with redundancy) the fundamental groups of the transverse manifolds at $\xi$. We illustrate the foregoing in the case of a $(1\cdot2\cdot2\cdot1)$-vertex.

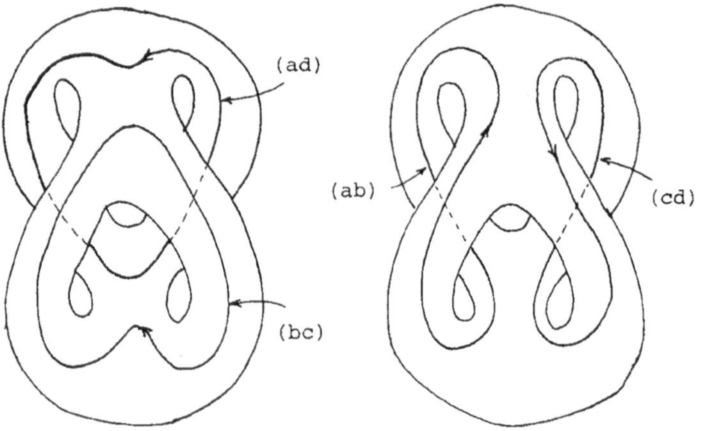


In all but the first case we've labelled the arcs of $C$ so that we can identify points on them as $\xi^{\pm}$, $\xi'^{\pm}$, $\xi^{0}$, $\xi^{1}$. We give a complete catalog of words in the arc labels of $A_{\xi}$ which label elements of $A_{\xi^*}$. (The $*$ superscript of $\xi^*$ and $c^*_{\xi}$ will denote one of $\pm$, $'\pm$, $0$, $1$.) The curves given are regularly homotopic, with fixed end-points, to arcs of $A_{\xi}$. Except in a neighborhood of a cusp, the regular homotopy that takes such a curve to the element of $A_{\xi^*}$ is constant on a neighborhood of the end points.

In the following catalog of elements of $A_{\xi^*}$, labelled by words $w$ in $A_{\xi}$ arc-labels we follow the convention that a point of $S(f)$ in the transverse manifold is denoted by $\cdot$ (or $\bigcirc$) if the transverse manifold is oriented $\odot$ (or $\circledcirc$). At non-vertex points this means that the arc of $S(f)$ points out of (or into) the plane of the page.

1) If $\xi \in \Sigma_1 - V$

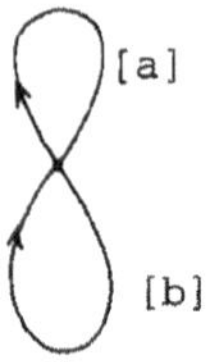

2) If $\xi$ is of type $(1\cdot2\cdot2\cdot1)$

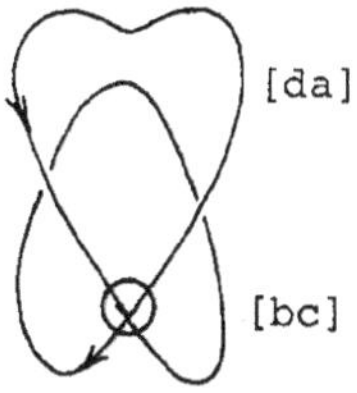


$q^{-1}(\xi^-)$

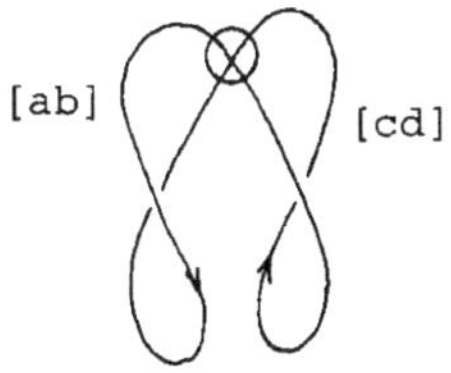


$q^{-1}(\xi'^+)$

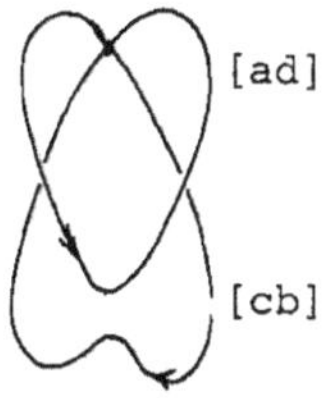


$q^{-1}(\xi^+)$

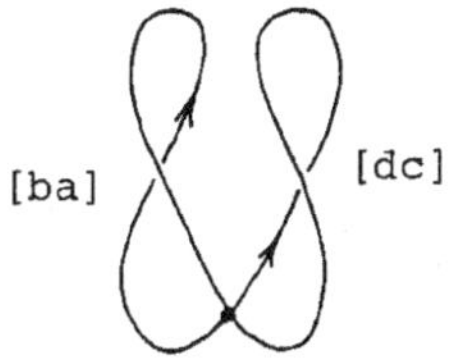


$q^{-1}(\xi'^-)$

3) If $\xi$ is of type $(1\cdot2\cdot2\cdot3)$

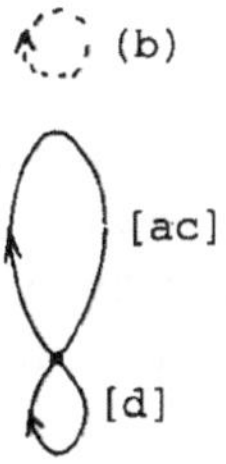


$q^{-1}(\xi^-)$

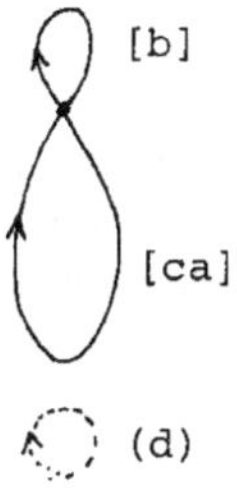


$q^{-1}(\xi'^+)$

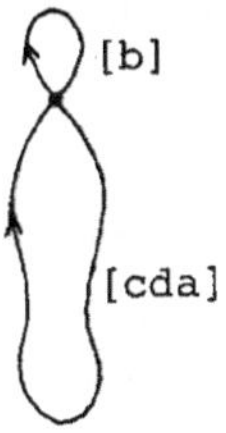


$q^{-1}(\xi^+)$

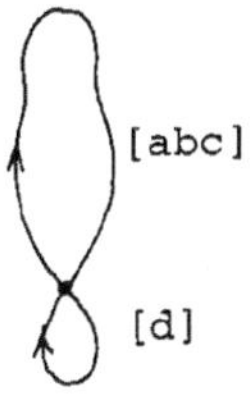


$q^{-1}(\xi'^-)$

4) If $\xi$ is a cusp

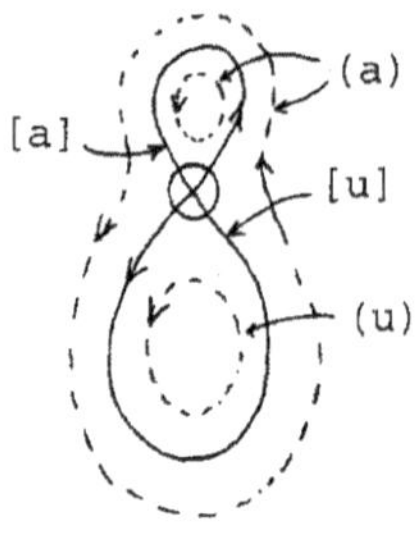


In case 3) we've dotted the curves (b) and (d) since they are not in $q^{-1}(\xi^-)$ or $q^{-1}(\xi'^+)$. In case 4) we've dotted in the two versions of (a) as well as (u) none of which are in $q^{-1}(\xi^1)$. Also in this case during the homotopy between the arc [a] in $q^{-1}(\xi^1)$ and [a] in $q^{-1}(\xi)$, $\xi$ the cusp, the angle between the tangent lines to the initial and terminal points of [a] decreases to zero as $\xi^1$ approaches $\xi$. If $v \in V_* \cap \bar{c} = v(c)$ for $c \in C_1$, and if $w$ is an arc-word in the arc-labels at $v$ and $[w]$ is regularly homotopic with fixed endpoints to an element $a \in A_\xi$ for $\xi \in c$ we say that $w$ or $[w]$ <u>represents</u> $a$ and we denote the word by $w_v(a)$. Similarly if $w$ is an arc-word in the arc-labels of $A_v$ and $[w]$ is one of the arcs in the preceding catalog there is a unique $c \in C_1$ with $v \in v(c)$ and an element $a \in A_\xi$ for $\xi \in c$ for which $w = w_v(a)$. We denote that element $a$ as $a[w]$. Similarly suppose $v \in V_* \cap \bar{R} = v(R)$ for $R \in \mathcal{R}$. If $w$ is a cycle-word in the arc labels of $A_v$ such that $(w)$ is regularly homotopic in the transverse manifold at $v$ to the fibre of the region $R \in \mathcal{R}$ we say that $w$ or $(w)$ <u>represents</u> $R$ and we denote the unique word in the arc-labels of $v$ as $w_v(R)$. Any cycle word $w$ gives rise to a closed curve $(w)$ in the transverse manifold to $v$ which represents a unique region $R \in \mathcal{R}$; we denote that region by $R(w)$. (I.e. $w_v(R(w)) = w$ and $R(w_v(R)) = R$.)

Let $(f,h) \in I_S$, we have $\sigma_h : A \longrightarrow \mathbb{R}$ defined on all $[a]$ for $a \in A_\xi$ for $\xi \in \Sigma_1$. We extend the definition to all arc-words in the arc labels of $A_\xi$ by $\sigma_h[w] = \mathrm{rot}(h\circ[w])$. Similarly $r_h : \mathcal{R} \longrightarrow \mathbb{Z}$ is extended to all cycle words in the arc-labels of $A_\xi$ as $r_h(w) = \mathrm{rot}(h\circ(w))$. Obviously:

$$\begin{cases} \sigma_h[a[w]] = \sigma_h[w] \\ \sigma_h[a] = \sigma_h[w_v(a)] \quad \text{for} \quad a \in A_\xi \quad \text{for} \quad \xi \in c \in C \quad \text{and} \quad v \in v(c) \end{cases}$$

and

$$\begin{cases} r_h(R(w)) = r_h(w) \\ r_h(R) = r_h(w_v(R)), \quad \text{for} \quad v \in v(R) \end{cases}$$

Just as we did in defining $\sigma_h : A \longrightarrow \mathbb{R}$, we write our extended $\sigma_h = \nu_h - \omega_h$ where $\nu_h$ is integer valued and $\omega_h$ takes values in $[-1/2,1/2)$ except if $v$ is a cusp and $a \in A_v$ in which case $\omega_h[a] = \frac{1}{2}\delta_h(v)$. Thus $2\pi\omega_h[a_1 \ldots a_n]$ is <u>the angle between the terminal direction of</u> $h \circ a_n$ <u>and the</u> initial direction of $h \circ a_1$. For any pair $a, b \in A_\xi$, $a \neq b$ define $u_h(a,b) \in [-1/2,1/2)$ so that $2\pi u_h(a,b)$ is <u>the angle between the terminal direction of</u> $h \circ a$ <u>and the initial direction of</u> $h \circ b$.

Exactly as in §3.6, since we are working with $(f,h) \in I_S$, we see that $\omega_h$ and $u_h$ depend only on the arcs in $A$ and $\delta_h$. The following relations are obvious:

($R_1$) Let $\xi \in c \in C_1$ and $v \in v(c) = V_* \cap \bar{c}$. Suppose $b \in A_\xi$ and $[a_1 \ldots a_n] = w_v(b)$ then

$$\nu_h(b) = \nu_h[a_1 \ldots a_n] = \sum_{i=1}^{n} \nu_h[a_i] + \sum_{i=1}^{n} (u_h(a_i,a_{i+1}) - w_h(a_i)),$$

where $a_{n+1} = a_1$.

($R_2$) If $v \in v(R) = V_* \cap \partial R$, then

$$r_h(R) = \nu_h[w_v(R)]$$

$$r_h(R) = \delta_h(c), \quad \text{if} \quad c \in C_0 \quad \text{is a part of} \quad \partial R.$$

<u>Note</u>. Since they only depend on $\delta_h$ we will write $u_{\delta_h}$ and $\omega_{\delta_h}$ for $u_h$ and $\omega_h$. Thus given $\delta \in \mathcal{D}$ we have the function $u_\delta$ and $\omega_\delta$.

By these two relations we see that $r_h$ and $\nu_h \mid A$ are determined by $\delta_h$ and $\nu_h \mid A^*$ with one exception. If $v$ is a cusp, the element $[u] \in A_\xi$ (for $\xi \in c \in C_1$ with $v \in v(c)$) is not represented by any arc-word at $v$. However in that case $\nu_h[u] = \delta_h(v)$. Thus we have:

<u>Proposition 0.</u> $r_h$ <u>and</u> $\nu_h \mid A$ <u>are completely determined by</u> $\delta_h$ <u>and</u> $\nu_h \mid A^*$.

Thus we see that for a given $\delta \in \mathcal{D}$ the functions $\nu : A \longrightarrow \mathbb{Z}$ that arise as $\nu_h$ for some $(h)_{\hat{\Sigma}}$ in $\mathcal{H}_{S,\Sigma}$ inject by restriction onto a subset of functions $\nu^* : A^* \longrightarrow \mathbb{Z}$ where the relation $(R_1)$ gives the recipe by which we must be able to extend $\nu^*$ to all of $A$.

<u>Definition.</u> For any $\delta \in \mathcal{D}$, we say that $\nu \in N_\delta \subseteq (C \cup A^*, \mathbb{Z})$, if extends uniquely to a map $\nu : C \cup A \longrightarrow \mathbb{Z}$ using $\delta$ and the formula $(R_1)$.

That is: Let $\xi \in c \in C$ and $v, v' \in v(c) = V_* \cap \bar{c}$ and $b \in A_\xi$. If $w_v(b) = [a_1 \ldots a_n]$ and $w_{v'}(b) = [a'_1 \ldots a'_m]$, then

$$\sum_{i=1}^{n} (\nu[a_i] + u_\delta(a_i, a_{i+1}) - w_\delta(a_i)) = \sum_{i=1}^{m} (\nu[a'_i] + u_\delta(a'_i, a'_{i+1}) - w_\delta(a'_i)).$$

The value of $\nu(b)$ is the common value of $\nu[w_v(b)]$ for all $v \in v(c)$.

<u>Proposition 1'.</u> <u>Let</u> $\delta \in \mathcal{D}$ <u>and</u> $\nu \in N_\delta$, <u>then there is an</u> $(h)_{\hat{\Sigma}} \in \mathcal{H}_{S,\Sigma}$ <u>such that</u> $\delta = \delta_h$ <u>and</u> $\nu = \nu_h$.

<u>Proof.</u> We construct $h$ in stages. In a neighborhood of $S(f)$, $h$ is fixed by the requirement that $(h)_{S(f)} = (h_{\delta,\nu})_{S(f)}$.

Now we proceed to define $h$ in a neighborhood $B(v)$ of each vertex $v \in V$. We work with the transverse manifolds, $T(v)$ and $\phi_v$ : $I \times T(v) \longrightarrow B(v)$ of the $C^2$PN at $v$. For all types of vertices we have our map defined on $\phi_v(I \times \mathbb{D})$ or $\phi_v(I \times \mathbb{D}_\pm)$ into $\mathbb{R}^2$. To extend $h$ we extend $k = h \circ \phi_v^{-1}$ to all of $I \times T(v)$. This is most easily described pictorially: the transverse manifolds at $v$ are:

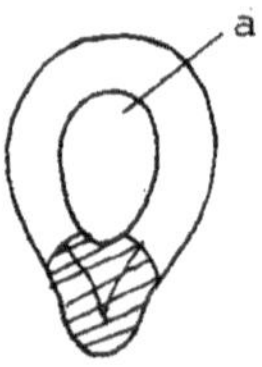


v, a cusp

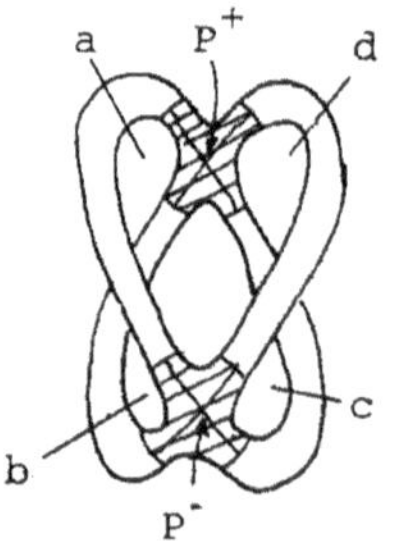


v, a (1·2·2·1) vertex

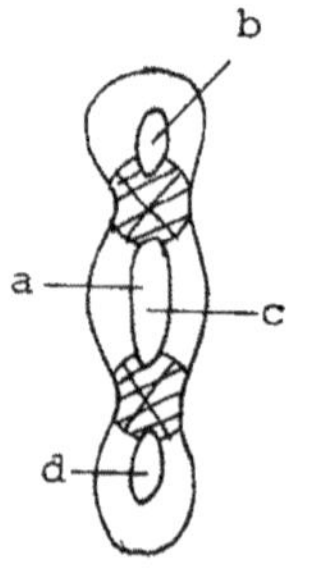


v, a (1·2·2·3) vertex

and the map $k$ is defined on $u \times$ (shaded regions) independent of $u$ (see §3.4). We extend $k$ to all of $u \times T(v)$ independent of $u$ so that $(f, k \circ \phi_v)$ is an immersion. We do this introducing the number of loops at each place in the figure as dictated by the values of $\nu$. We illustrate with an example at $(1\cdot 2\cdot 2\cdot 1)$ point. Suppose $\nu[a] = 1$, $\nu[b] = -3$, $\nu[c] = 1$, $\nu[d] = 0$. Suppose further than $\delta(c^+) = 1$ and $\delta(c^-) = -1$ where $P^+ \in c^+$ and $P^- \in c^-$. In that case $k(T(v))$ looks like:

If $c \in C_1$ is closed we define $k$ on $I_\pm \times T(c)$ by defining it on $I_+ \times T(c)$ exactly as we did for neighborhoods of the vertices and then extend the definition to $I_- \times T(c)$ as $k \circ \phi_{c,+}^{-1} \circ \phi_{c,-}$ (see §2.3). This gives us our definition of $k$ on neighborhoods of $v^c = q^{-1}(v_c) \cap S(f)$. Thus we have $h$ defined on neighborhoods $B(v)$ for all $v \in V_*$. In order to extend $h$ to a neighborhood of $q^{-1}(c)$ for all $c \in C_1$, we work in $I \times T(c)$, of the $C^2$PN of $f$ along $c$. So far we have $k$ defined on $(I \times \mathbb{D}) \cup (I_\pm \times T(c))$. By definition of $h_{\delta,\nu}$ (see §3.5) for each $u \in I$, $k \mid (u \times \mathbb{D})$ is an embedding. For all $u \in (I - I_\pm)$, $k(u,z) = R(u)J_-(z)$ where $R : I \longrightarrow O(2)$; $R \mid I_-$ = identity on $\mathbb{R}^2$ and $R \mid I_+$ is rotation through $2\pi\tau(c)$. Define $k : I - I_\pm^\circ \times T(c) \longrightarrow \mathbb{R}^2 : (u,z) \longrightarrow R(u)k(-1+\alpha, z)$. This together with $k \mid I_- \times T(c)$ can be smoothed to give the desired extension of $k$ to $(I - I_+^\circ) \times T(c)$. However the two maps defined on $(1-\alpha) \times T(c)$ may only agree on $(1-\alpha) \times \mathbb{D}$. However we can easily redefine $k$ on $[1-\alpha, 1-\alpha'] \times T(c)$, $0 < \alpha' < \alpha$, using Proposition 3 of §3.6 so as to obtain the desired extension of $k$ to all of $I \times T(c)$. //

In §3.6 we introduced the map $\lambda : H_{S,\Sigma} \longrightarrow \mathcal{D} \times (C \cup A, \mathbb{Z})$ and set $L_\Sigma = \lambda(H_{S,\Sigma})$. We have thus proven:

Proposition 1. $L_\Sigma = \cup\{(\delta \times N_\delta),\ \delta \in \mathcal{D}\}$.

Definition. For any $\delta \in \mathcal{D}$ we let $N'_\delta$ be those maps $\nu \in N_\delta$ which define unique maps $\tilde{r} = \tilde{r}(\delta,\nu) : R \longrightarrow \mathbb{Z}$ using the formula $(R_1)$ and $(R_2)$.

That is if $v, v' \in v(R) = V_* \cap \partial R$, then

$$\nu[w_v(R)] = \nu[w_{v'}(R)].$$

To compute these integers we must use $(R_1)$. This value of $\tilde{r}(R)$ is the common value of all $\nu[w_v(R)]$ for all $v \in v(R)$.

If we let $L'_\Sigma = \{\delta \times N'_\delta \mid \delta \in \mathcal{D}\}$ then we know that if $\lambda : I_S \longrightarrow \mathcal{D} \times (C \cup A, \mathbb{Z}) : (f,h) \longrightarrow (\delta_h, \nu_h)$ then $\lambda(I_S) \subseteq L'_\Sigma$.

Although this image $\lambda(I_S)$ is generally a proper subset of $L'_\Sigma$ the functions $\tilde{r}(\delta,\nu)$ defined using $(R_1)$ and $(R_2)$ share one property with $r_h$ for $(f,h) \in I_S$:

Proposition 2. $\tilde{r}(\delta,\nu) \in \mathrm{Rot}(\vartheta_\delta)$ for $\nu \in N'_\delta$ where $\vartheta_\delta$ is the orientation defined by $\vartheta_\delta(c) = (-1)^i\delta(c)$ for $c \in C_i$.

Proof. This only needs to be checked in a neighborhood of a point of $c \in C_1$ since if $R$ has an element $c \in C_0$ as part of its boundary we know that

$$\tilde{r}(\delta,\nu)(R) = \delta(c) \quad \text{as required.}$$

Now consider three regions that abut an element $c \in C_1$, one stem and two arm regions whose cycle-words, in arc labels in $A_\xi$ for $\xi \in c$, are respectively (ab), (a) and (b). Letting $\tilde{r} = \tilde{r}(\delta,\nu)$ we must check:

$$\tilde{r}(a) + \tilde{r}(b) - \tilde{r}(ab) = -\delta.$$

Using $(R_1)$ and $(R_2)$ we must check that

$$\omega_\delta(a) + \omega_\delta(b) - u_\delta(a,b) - u_\delta(b,a) = -\delta.$$

To evaluate the left hand side of this equation we must consider a map $h$ whose germ $(h)_{\hat{\Sigma}} \in \mathcal{H}_{S,\Sigma}$ and such that $\delta = \delta_h$. Except for a rotation this specifies the germ of $h$ at any point of $S(f)$. We examine the image under such an $h$ of a neighborhood of $q^{-1}(\xi) \cap S(f)$ in the transverse manifold:

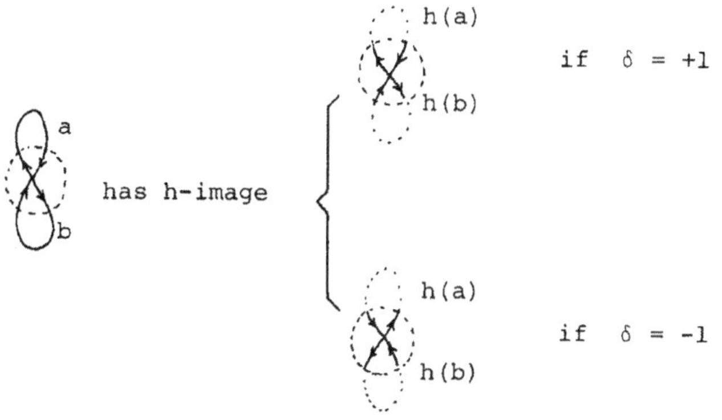


Thus $\omega_\delta(a) = \omega_\delta(b) = -\delta/4 = -u_\delta(a,b) = -u_\delta(b,a)$. //

Remark. For each $\delta \in \mathcal{D}$, $N_\delta$ and $N'_\delta$ are submodules of $(C \cup A, \mathbb{Z})$.

## 3.8 The additional relations on $\{\delta,\nu\}$ arising from the global lift of f

As has already been remarked in the Proposition of §3.1, the map that takes $(f,h) \in I(B(R),\mathbb{R}^4)$ to the homomorphism $(P_{Th})_* : H_1(B(R)) \longrightarrow \mathbb{Z}$ gives a 1:1 correspondence between the f-regular homotopy classes, $[I(B(R),\mathbb{R}^4)]$ and $H^1(B(R),\mathbb{Z})$. Having chosen a trivialization $\phi_R : R \times S^1 \longrightarrow B(R)$, we write $(P_{Th}\circ\phi_R)_* : H_1(R) \oplus H_1(S^1) \longrightarrow \mathbb{Z}$ as the sum of two homomorphisms $s_h(R) : H_1(R) \longrightarrow \mathbb{Z}$ and $r_h(R) : H_1(\mathbb{S}^1) \longrightarrow \mathbb{Z}$. Thus the f-fibre regular homotopy classes $[I(B(R),\mathbb{R}^4)]$ are classified by two homomorphisms: $s_h(R)$ and $r_h(R)$, where, as usual, we identify $r_h(R)$ with its value on the orientation generator of $H_1(\mathbb{S}^1)$.

In §3.7 we defined a subset $L'_\Sigma = \cup\{\delta \times N'_\delta \mid \delta \in \mathcal{D}\} \subseteq \mathcal{D} \times (C \cup A^*,\mathbb{Z})$ such that $(\delta,\nu) \in L'_\Sigma$ iff 1) $\nu$ extends uniquely to an element of $(C \cup A,\mathbb{Z})$ as prescribed by the equations $R_1)$ given there and 2) there is a function $\tilde{r}(\delta,\nu) \in \mathbb{Z}^{\mathcal{R}}$ linear in $\delta$ and $\nu$ prescribed in $R_1)$ and $R_2)$ such that $\tilde{r}\circ\lambda(f,h) = r_h$ for $(f,h) \in I_S$.

These conditions of membership in $L'_\Sigma$ are given as the vanishing of a set of linear expressions in the values of $\delta$ and $\nu$. We have then a commutative square:

$$\begin{array}{ccc} [I_S]_S & \xrightarrow{\ \lambda\ } & L'_\Sigma \\ \Big\downarrow & \searrow^{r} & \Big\downarrow \tilde{r} \\ [I(B(R),\mathbb{R}^4)] & \xrightarrow[r]{} & \mathbb{Z} \end{array}$$

where the unlabelled vertical map is the obvious projection.

In this section we show that there is an analogous function $\tilde{s}$ : $L'_\Sigma \longrightarrow \underset{R\in R}{\times} (H^1(\partial R,\mathbb{Z})$, linear in $\delta$ and $\nu$ such that the following diagram commutes:

$$\begin{array}{ccc} [I_S]_S & \xrightarrow{\ \lambda\ } & L'_\Sigma \\ \Big| & & \Big\downarrow \tilde{s} \\ \Big| & \searrow^{s} & H^1(\partial R,\mathbb{Z}) \\ \Big| & & \Big\downarrow i^* \\ \downarrow & & \\ [I(B(R),\mathbb{R}^4] & \xrightarrow[s]{} & H^1(R,\mathbb{Z}) \end{array}$$

where $H^1(\partial R,\mathbb{Z}) = \underset{R\in R}{\times} H^1(\partial R,\mathbb{Z})$ and $H^1(R,\mathbb{Z}) = \underset{R\in R}{\times} H^1(R,\mathbb{Z})$.

Let $L = \tilde{s}^{-1}(i^*(H^1(R,\mathbb{Z})))$. As usual, we write

$$L = \cup\{\delta \times N^*_\delta,\ \delta \in D\} \quad \text{where} \quad N^*_\delta \subseteq N'_\delta \subseteq N_\delta .$$

For each $R \in R$, let $d_j(R) \in H_1(\partial R)$, $j = 1,\dots,b(R)$ be the class of the oriented component $\beta_j(R)$ of $\partial R$. Their images $b_j(R) = i_* d_j(R)$ satisfy a single relation: $\Sigma\, b_j(R) = 0$. Thus, an element $\sigma \in H^1(\partial R,\mathbb{Z})$ is in $i^* H^1(R,\mathbb{Z})$ iff for each $R \in R$, $\sum_{j=1}^{n(R)} \sigma(R)(d_j(R)) = 0$.

In addition to the generators $\{b_j(R),\ j = 1,\dots,n(R)\}$, there are $2m(R)$ free generators $\{k_i(R), k'_i(R),\ i = 1,\dots,m(R)\}$ of $H_1(R)$, representing the handles of $R$.

Let $Z = \underset{R\in R}{\times} (\mathbb{Z}^{2m(R)})$, and define $\zeta : [I_S]_S \longrightarrow Z : [f,h] \longrightarrow \zeta_h$ where $\zeta_h(R) = (s_h(R)(k_i(R)),\ s_h(R)(k'_i(R)),\ i = 1,\dots,m(R))$.

We put all of our maps together:

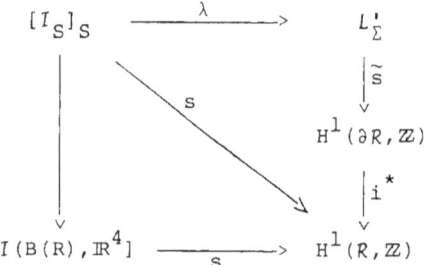

$$\begin{array}{ccc}
[\mathcal{I}_S]_S & \xrightarrow{(\lambda\times\zeta)} & \mathcal{L}\times\mathcal{Z} \\
 & & \big\downarrow \tilde{r}\times\tilde{s}\times 1 \\
\big\downarrow & \searrow^{r,s} & \mathbb{Z}^{\mathcal{R}}\times H^1(\partial\mathcal{R},\mathbb{Z})\times\mathcal{Z} \\
 & & \big\downarrow 1\times i^*\times k \\
[\mathcal{I}(B(\mathcal{R}),\mathbb{R}^4)] & \xleftrightarrow{(r,s)} & \mathbb{Z}^{\mathcal{R}}\times H^1(\mathcal{R},\mathbb{Z})\cong H^1(B(\mathcal{R}),\mathbb{Z})
\end{array}$$

where $k(\sigma)(R) = \{\sigma(R)(k_i(R)),\sigma(R)(k_i'(R)),\ i = 1,\dots,m(R)\}$ for any $\sigma \in H^1(\mathcal{R},\mathbb{Z})$. Obviously the map $(1\times i^*\times k)$ is injective. Assuming the existence of the map $\tilde{s}$ we prove:

<u>Proposition</u>. <u>The map</u> $(\lambda\times\zeta) : [\mathcal{I}_S]_S \longrightarrow \mathcal{L}\times\mathcal{Z}$ <u>is surjective</u>.

<u>Proof</u>. For any $\nu \in \mathcal{N}_\delta^* \subseteq \mathcal{N}_\delta$ we know by Proposition 1 of §3.7 that there is an $(h') \in \mathcal{H}_{S,\Sigma}$ with $(\delta,\nu) = (\delta_{h'},\nu_{h'})$. Also by Proposition 2 of §3.7 since $\nu \in \mathcal{N}_\delta^* \subseteq \mathcal{N}_\delta'$, we can construct $\tilde{r}(\delta,\nu) \in \mathrm{Rot}(\vartheta_\delta)$. This together with $\tilde{s}(\delta,\nu)$ and any element $\zeta \in \mathcal{Z}$ determines an element of $\mathbb{Z}^{\mathcal{R}}\times H^1(\mathcal{R},\mathbb{Z})$ which in turn determines an element $[(f,h'')] \in [\mathcal{I}(B(\mathcal{R}),\mathbb{R}^4)]$. For any representative map $(f,h'')$ we have $r_{h''} = \tilde{r}(\delta,\nu)$ and $(i^*\times k)s_{h''} = (\tilde{s}(\delta,\nu);\zeta)$.

Choosing representative maps $h' : B(\Sigma) \longrightarrow \mathbb{R}^2$ and $h'' : M - \hat{\Sigma} \longrightarrow \mathbb{R}^2$, we must see that we can glue these maps together to get a map $h : M \longrightarrow \mathbb{R}^2$ such that $(f,h) \in \mathcal{I}_S$ and $(\lambda_h,\zeta_h) = ((\delta,\nu),\zeta)$.

Since the homomorphisms, $(P_{Th''})_*$ and $(P_{Th'})_*$ mapping $H_1(\partial B(\Sigma))$ to $\mathbb{Z}$ are determined in the same way by the given $\delta$ and $\nu$ we know that they are equal. Hence the f-fibre homotopy classes of $h' \mid \partial B(\Sigma)$ and $h'' \mid \partial B(\Sigma)$ are the same [E-G Theorem 1.2.5]. So we can connect the two maps in a little collar neighborhood of $\partial B(\Sigma)$ in which we extend the f-fibre regular homotopy beginning with $h'$ on one boundary component of the collar neighborhood and ending with $h''$ on the other. The resulting map $h$ agrees with $h'$ in a neighborhood of $S(f)$ (in fact on a neighborhood of $\hat{\Sigma}$) so that $(\delta_h,\nu_h) = (\delta_{h'},\nu_{h'})$. The map $h$ agrees with $h''$ on the complement of a neighborhood of $\hat{\Sigma}$ so that the values of $s_h$ and $s_{h''}$ agree on the handle generators $k_i(R)$ and $k_i'(R)$ in $H_1(R)$. Thus $\zeta = \zeta_h$. //

The map $\tilde{s} : \mathcal{L}_\Sigma' \longrightarrow H^1(\partial\mathcal{R},\mathbb{Z})$ that we want must satisfy: For every $R \in \mathcal{R}$, and $(f,h) \in \mathcal{I}_S$,

$$s_h(R)(b_j(R)) = \tilde{s}(\delta_h,\nu_h)(R)(d_j(R)),\quad j = 1,\dots,n(R).$$

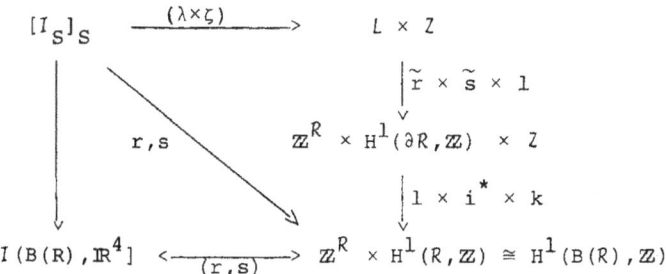

For each $R$ and $b_j(R)$ having determined a linear expression in the values of $\delta_h$ and $\nu_h$ which gives $s_h(R)(b_j(R))$, we define $\tilde{s}(\delta,\nu)(R)(d_j(R))$ as that expression in the values of $\delta$ and $\nu$.

Once $\tilde{s}$ is defined, we define $L \subseteq L'_\Sigma$ as those $(\delta,\nu) \in L'_\Sigma$ such that for all $R \in \mathcal{R}$,

$$\sum_{j=1}^{n(R)} \tilde{s}(\delta,\nu)(R)(d_j(R)) = 0.$$

To find the linear expression in the values of $\delta_h$ and $\nu_h$ which gives the value of $s_h(R)(b_j(R))$, we deform the map $\phi_R : \partial R \times 1 \longrightarrow B(R)$ so that it consists of a sequence of arcs: some in the fibres above points of $R$ near the vertices of $\partial R$ and others close to the arcs of $S(f)$ which join a pair of q-pre-images of those vertices. For any component $\beta_j$ of $\partial R$, we obtain arcs close to elements of $A^*$ and others close to arcs of $S(f)$ above elements of $C$. The contributions to the value of $s_h(R)(b_j(R))$ coming from the arcs close to those of $A^*$ are essentially the values of $\sigma_h$ on those elements of $A^*$ and coming from the arcs close to those of $S(f)$ are essentially the values of $\tau_h$ on the corresponding elements of $C$. To eliminate the "essentially 's" in the preceding sentence we must add the rotation numbers of $h$ composed with arcs that smooth the transitions between the arcs near those of $A^*$ and $S(f)$. These transition arcs are all near $S(f)$ and are independent of $h$ and the computation of the rotation numbers only involves $h$ as a $\delta_h$ factor. Since both $\sigma_h$ and $\tau_h$ are expressible in terms of $\delta_h$ and $\nu_h$, we have found an expression for $s_h(R)(b_j(R))$ in terms of values of $\delta_h$ and $\nu_h$ as required.

The rest of this section is devoted to the detailed description of the determination of $\tilde{s}$.

We will work entirely in a collar neighborhood of $\partial R$, $B(\Sigma) \cap B(R)$ over $N(\Sigma) \cap R$, (see §1.6). Our first step is to choose a curve (perhaps with many components), $\underline{\partial R}$, in $R$ deforming $\partial R$ close to $\Sigma$. In particular suppose an oriented component $\beta$ of $\partial R$ is given by sequence of arcs $c_0c_1\ldots c_k$ where $c_j$ are arcs in $\Sigma - V$ oriented as parts of $\partial R$. As a chain, $\Sigma c_j = \Sigma n(R,c)c$ where $c \in C$, arcs in the component $\beta$ of $\partial R$ with the standard orientation (§1.5). The arc $c_{j-1}$ joins a vertex $v_{j-1}$ to a vertex $v_j$ $(v_{k+1} = v_0)$; we delay discussing the case of a boundary component without vertices. The corresponding component $\underline{\beta}$ of $\underline{\partial R}$ consists of curves $\underline{c}_j$ close to $c_j$ in $N(c_j)$ joining points $\underline{v}_{j-1}$ to $\underline{v}_j$ where $\underline{v}_\cdot \in N(v_\cdot)$. The following essentially complete collection of figures will be used as a reference for the descriptions of the lift of $\underline{\partial R}$ to $B(\Sigma) \cap B(R)$.

For v a (1·2·2·1)-vertex

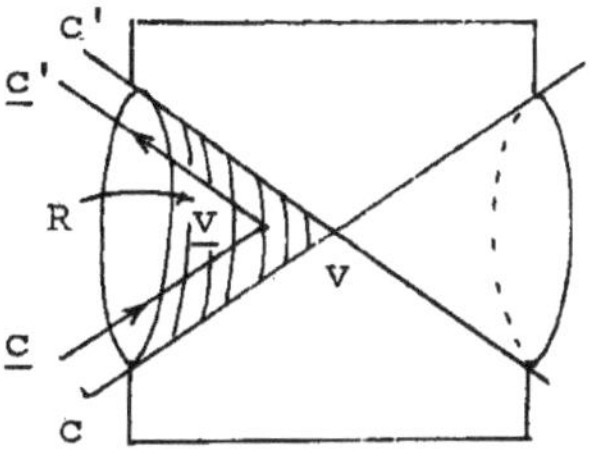


(i)

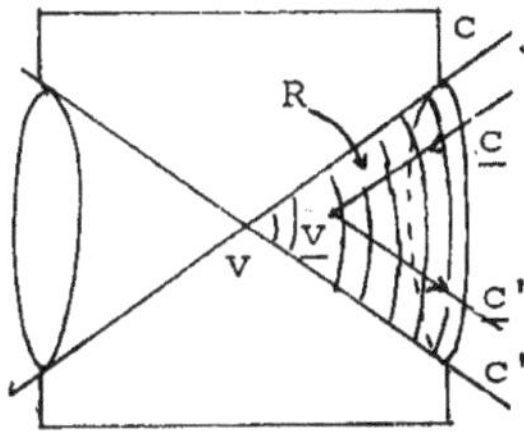


(ii)

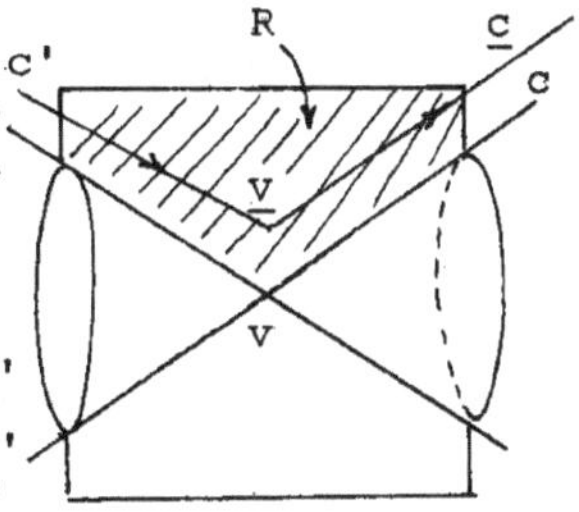


(iii)

For v a (1·2·2·3)-vertex

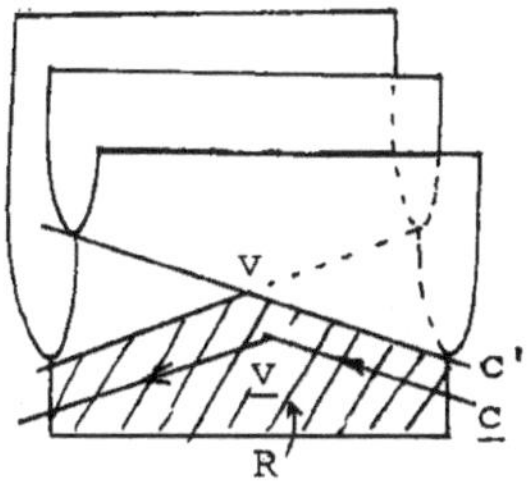


(iv)

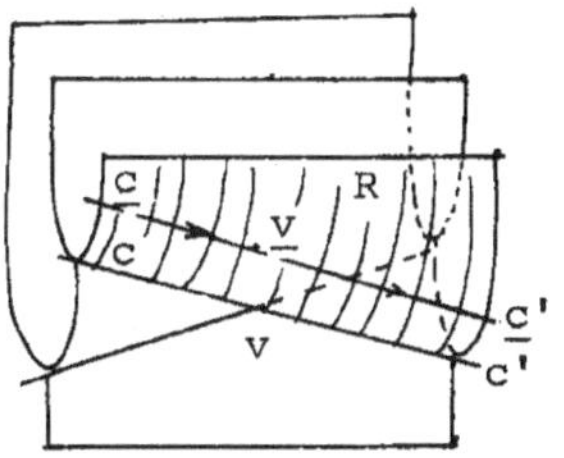


(v)

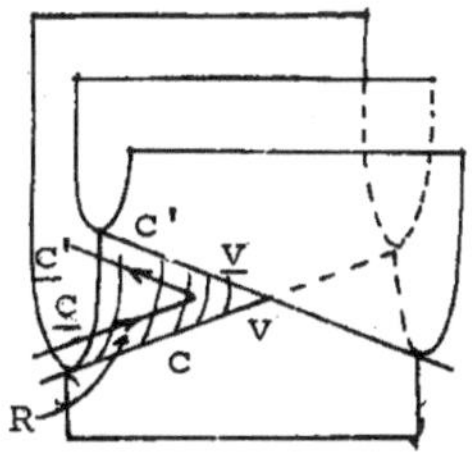


(vi)

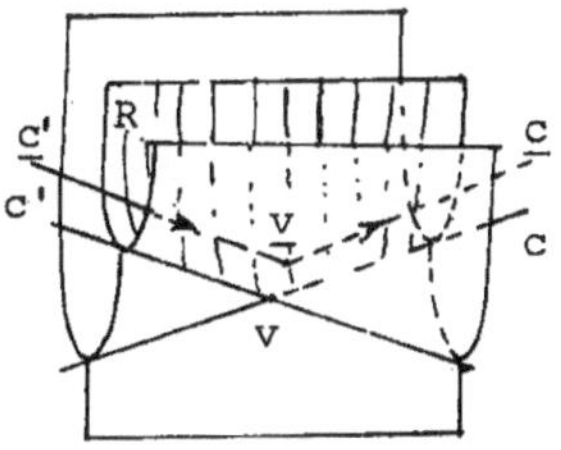


(vii)

For v a cusp

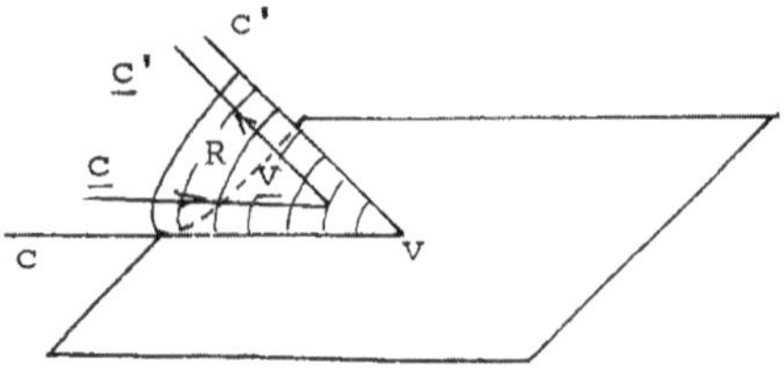


(viii)

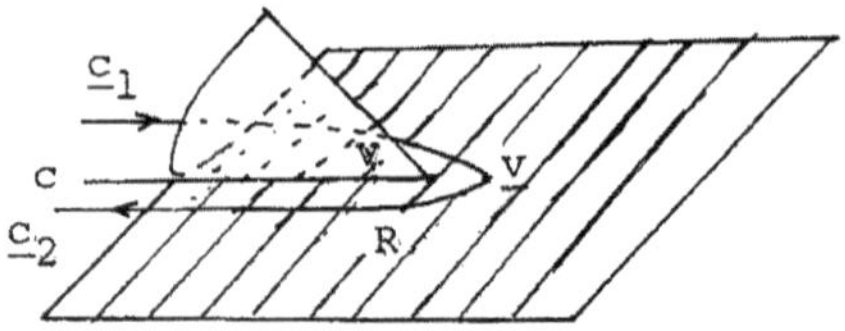


(ix)

The neighborhoods of v pictured above are $N(v)$. For each $v \in V$, $B(v) = q^{-1}(N(v))$ which we know to be a product $I \times T(v)$ where $T(v)$ is a transverse manifold to v. (See §1.6.) In this section if $\underline{v} \in N(v)$, the transverse manifold containing $q^{-1}(\underline{v})$ is the image of $t \times T(v)$ in $B(v)$ which contains $q^{-1}(\underline{v})$.

We now describe the method by which we lift the arcs $\underline{c}$ to $B(c) \cap B(R)$ for $c \in \mathcal{C}$.

Let $(\phi_c, \psi_c, g_c)$ be the $C^2$PN of f along c (see §2.3). Since $N_{\overline{f}|c}$ is trivial, we may assume that there is a diffeomorphism, $\gamma_c$ : $c \times \mathbb{R} \longleftarrow N_{\overline{f}|c}$ such that if $\sigma : I \longrightarrow c$ is the identification map we have the commutative diagram:

$$\begin{array}{ccc} I \times J & \xrightarrow{\psi_c} & \tilde{N}_c \subseteq N_{\overline{f}|c} \\ \sigma \times 1 \downarrow & & \downarrow \gamma_c \\ c \times J & \hookrightarrow & c \times \mathbb{R} \end{array}$$

Let $p_c : B(c) \longrightarrow \mathbb{R}$ be the composition $\mathrm{proj}_{\mathbb{R}} \circ \gamma_c \circ \tilde{f}$, and let $\tilde{c} = q^{-1}(c) \cap S(f)$.

<u>Definition</u>. In $B(c)$, let $X(c)$ be the set of points on the stream lines of grad $p_c$ which either emanate from or tend toward points of $\tilde{c}$ including points of $\tilde{c}$. For $c \in \mathcal{C}_0$, $X(c) = B(c)$, whereas for $c \in \mathcal{C}_1$, $X(c)$ meets every fibre of $B(c)$, i.e. every transverse manifold to c, in a pair of lines intersecting at a point of $\tilde{c}$.

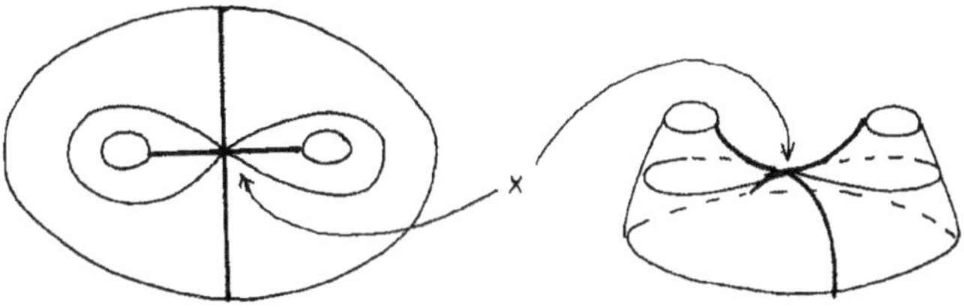


$X(c) \cap B(c)_x$

Given any curve in $N(c) - c$ for $c \in C$, we lift it to $B(c) - \tilde{c}$ as follows: (Recall $N(c) = q(B(c))$.)

1. For $c \in C_0$, $B(c)$ is also trivial so that we may assume that there is an embedding $\beta_c : B(c) \longrightarrow c \times \mathbb{R}^2$ so that the following diagram commutes.

$$\begin{array}{ccccccc}
I \times T(c) & \xrightarrow{\phi_c} & B(c) & \xrightarrow{\beta_c} & c \times \mathbb{R}^2 & \ni & (\xi, z) \\
\downarrow{\scriptstyle 1 \times g_c} & & \downarrow{\scriptstyle f} & & \downarrow & & \downarrow \\
I \times J & \xrightarrow{\psi_c} & N_{\bar{f}|c} & \xrightarrow{\gamma_c} & c \times \mathbb{R} & \ni & (\xi, |z|^2)
\end{array}$$

In this case $N(c)$ can be identified with $\gamma_c^{-1}(c \times [0,\infty))$ which identifies $f$ with $q$. Thus letting $\mathbb{R}^+ = (0,\infty)$ we have:

$$\begin{array}{ccc}
B(c) - \tilde{c} & \xrightarrow{\beta_c} & c \times \mathbb{R}^+ \times S^1 \\
\downarrow{\scriptstyle q} & & \downarrow \\
N(c) - c & \xrightarrow{\gamma_c} & c \times \mathbb{R}^+
\end{array}$$

We define a section of $q$ by choosing an element $z_0 \in S^1 \subseteq \mathbb{C}$ and mapping $c \times \mathbb{R}^+$ to $c \times \mathbb{R}^+ \times z_0$. For convenience we choose $z_0 = 1$.

2. For $c \in C_1$, the restriction of $q$ maps $X(c) - \tilde{c}$ onto $N(c) - c$ as a covering space. Each component of $N(c) - c$ is either singly or doubly covered. The fibre of $N(c)$ is a $Y$, the vertex of which is a point of $c$. The stem of $Y$ is double-covered and each of the arms of $Y$ is singly covered. Thus any curve in $N(c) - c$ has either a unique lift to $X(c) - \tilde{c}$ if the curve is an arm component of $N(c) - c$ or has two lifts if the curve is in the stem component of $N(c) - c$.

Notice that $N(c) - c$ has either three components, one stem and two arms if $N(c)$ is a trivial Y-bundle; or two components, one stem and one arm if $N(c)$ is the non-trivial Y bundle described in §1.6.

We now deform the curve $\phi_R(\underline{\beta})$: During the deformation points slide only in their own q-fibres. We deform the curves in such a way that the curve above each $\underline{c}_i$ will be one of the lifts just described, $\hat{c}_i$. As a result of this deformation of $\phi_R(\underline{c}_i)$ into $\hat{c}_i$, the end point of $\phi_R(\underline{c}_{i-1})$ and the initial point of $\phi_R(\underline{c}_i)$ trace out a curve $\hat{v}_{i-1}$ in the q-fibre of $\underline{v}_{i-1}$. As a result of deforming $\phi_R(\underline{\beta})$ in this way, we obtain a closed curve which we denote by $\hat{c}_0\hat{v}_0 \ldots \hat{c}_k\hat{v}_k$. In the following figures, we indicate the possible $\hat{v}$-arcs. To make the figures more meaningful we draw the transverse manifolds containing $q^{-1}(\underline{v})$ and indicate on this q-figure the points of entry and egress of the $\hat{c}_i$ by $\cdot$ and $O$ respectively. In each figure we indicate one of the possible pairs of $\hat{c}_i$-arcs. Besides the initial and final point possibilities we know nothing more about the arc $\hat{v}$. It may traverse the q-fibre in either sense and it may go around the fibre completely many times. These figures are numbered to agree and use the same notations as the figures showing $\underline{c}$, $\underline{c}'$ and $\underline{v}$, at the beginning of this section.

The last two figures require some explanation. (ixa) is the transverse manifold through the cusp, and (ixb) is above $\underline{v}$ beyond the cusp. The pattern of q-fibres on all of the transverse manifolds beyond the cusp all look alike and the three points (two egress and one entry) coalesce to one above $\underline{v}$.

We use the notation for arc-labels and words introduced in §3.7. If $v$ is any vertex on the boundary of $R \in \mathcal{R}$ there is a cycle word $w_v(R)$, unique up to cyclic permutations, representing $R$. Any arc $\hat{v}$ which is on $q^{-1}(\underline{v})$ which connects the end of one arc $\hat{c}$ to the beginning of another $\hat{c}'$ can also be described by a word, $w_v(\hat{v})$ in the arc labels of $A_v$. This word begins at an arc-label $a$, whose arc originates at the point of $q^{-1}(v) \cap S(f)$ which is near the end point of $c$ proceeds through the arc labels in $w_v(R)$ till meeting an arc label $z$ whose arc ends at a point of $q^{-1}(v) \cap S(f)$ which is near the initial point of $\hat{c}'$. This sequence of arc labels is followed by a number, $(\pm k)$, of cycles $w_v(R)$ ending with the arc-label, $z$.

We list below the words $w_v(\hat{v})$ that can arise. The first one listed is the one using the entering and leaving lifted arcs of $C$ in the figures. In brackets we give words belonging to $\hat{v}$ similar to the ones in the figures but lying on q-fibres over $\underline{v}$ in other elements of $\mathcal{R}$ abutting the same vertex.

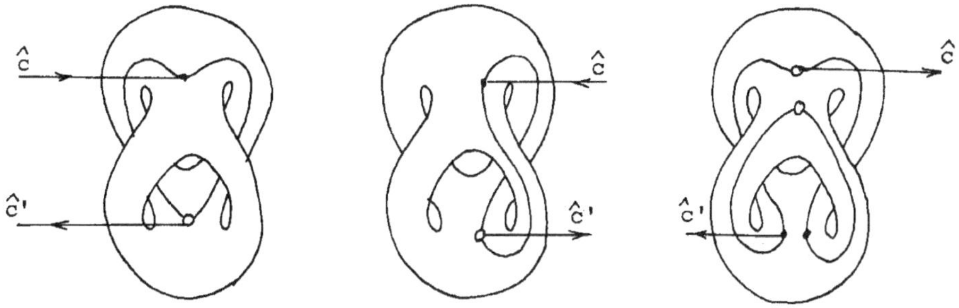


(i) (ii) (iii)

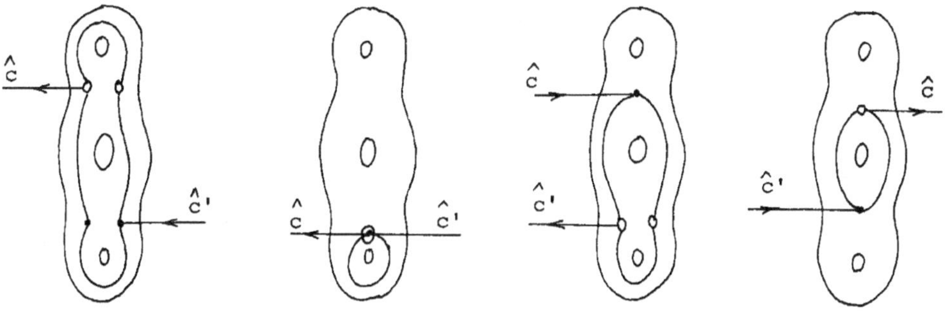


(iv) (v) (vi) (vii)

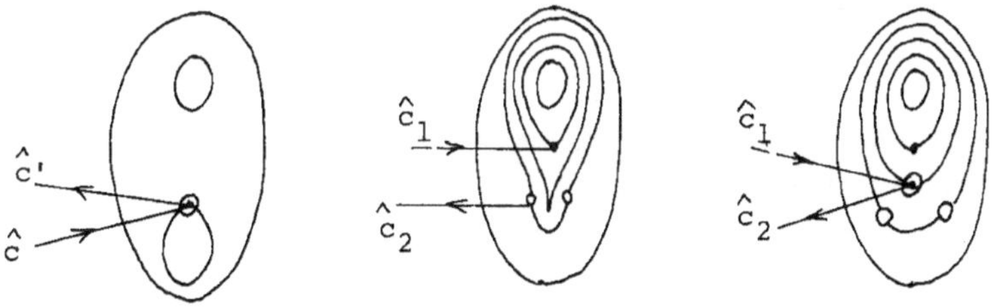


(viii) (ixa) (ixb)

(i) $a(da)^{\pm k}$, $\{c(bc)^{\pm k}\}$
(ii) $c(dc)^{\pm k}$, $\{a(ba)^{\pm k}\}$
(iii) $b(cdab)^{\pm k}$; $bcd(abcd)^{\pm k}$, $d(abcd)^{\pm k}$, $dab(cdab)^{\pm k}$, $\{b(adcb)^{\pm k}$, $bad(cbad)^{\pm k}$, $d(cbad)^{\pm k}$, $dcb(adcb)^{\pm k}\}$
(iv) $d(abcd)^{\pm k}$, $da(bcda)^{\pm k}$, $dab(cdab)^{\pm k}$, $a(bcda)^{\pm k}$, $ab(cdab)^{\pm k}$
(v) $d^{\pm k}$, $\{b^{\pm k}\}$
(vi) $c(dac)^{\pm k}$, $cd(acd)^{\pm k}$, $\{a(bca)^{\pm k}$, $ab(cab)^{\pm k}\}$
(vii) $a(ca)^{\pm k}$
(viii) $u^{\pm k}$ ($u$ is the symbol for the whole oriented fibre, $q^{-1}(\underline{v})$)
(ix) $a^{\pm k}$ ($a$ is the symbol for the whole oriented fibre $q^{-1}(\underline{v})$.

In case (ix) of lifting a curve around the cusp, the $\underline{c}_1$ arc is lifted as usual until its lift meets the transverse manifold through the cusp. After that it travels to a point on a non-singular q-fibre in a transverse manifold above a point beyond the cusp, loops around the fibre as many times as it needs to and then returns as $\hat{c}_2$. The lift $\hat{c}_2$ of $\underline{c}_2$ is similar to the lift $\hat{c}_1$ of $\underline{c}_1$. In making the turn from $\hat{c}_1$ to $\hat{c}_2$, the tangents to the q-fibres encountered turn through $\pm\pi/2$ in addition to the $\pm k(2\pi)$ in the a-fibre loops.

For the non-cusp cases we describe how to construct the arc $\hat{v}$ from the word $w_v(\hat{v})$. Let $a_0 \dots a_n$ be the cycle word in the arc labels of $A_v$ representing $R$; the closed arc in the transverse manifold $T(v)$ representing $R$ is then $(a_0 \dots a_n)$. The word $w_v(\hat{v})$ also consists of a sequence $a_j a_{j+1} \dots a_r$ in increasing or decreasing order. The arc, $\hat{v}$, will be denoted by $\langle a_j \dots a_r \rangle$ which is the same as $[a_j \dots a_r]$ in the notation of §3.7 except that it begins with the last half of the transitional quarter circle arc between $a_{j-1}$ and $a_j$ and ends with the first half of the transitional quarter circle arc between $a_r$ and $a_{r+1}$. (As usual $a_{-1} = a_n$ and $a_{n+1} = a_0$.)

We give an example. Suppose $v$ is of type $1\cdot 2\cdot 2\cdot 1$ the if $w(R) = adcb$, we illustrate $q^{-1}(v)$, $(adcb)$ and $\langle dcb \rangle$

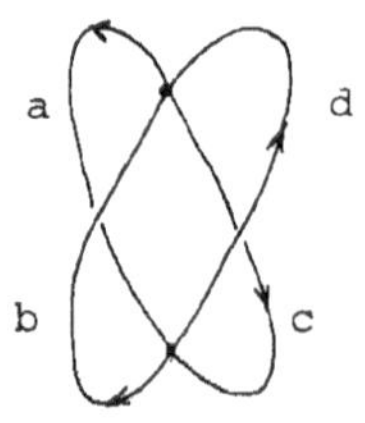

$q^{-1}(v)$

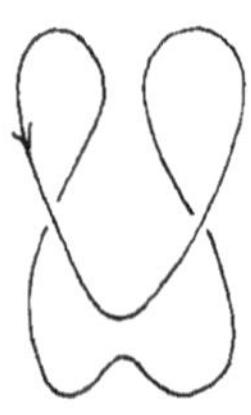
(adcb)

<dcb>

If $c \in C$ is closed and contained in the boundary of $R$, we consider $c$ as joining $v_c \in c$ with itself. If $B(c)$ is trivial, then $(N(c)-c) \cap R$ is diffeomorphic to $c \times (0,1)$ so we can choose a $\underline{v}_c$ near $v_c$ and a $\underline{c}$ which is carried to $c \times \varepsilon$ by the diffeomorphism (where $\underline{v}_c = v_c \times \varepsilon$.)

If $B(c)$ is not trivial, there are two cases: $(N(c)-C) \cap R$ is diffeomorphic to $c \times (0,1)$ in which case we choose $\underline{v}_c$ and $\underline{c}$ as above. In the other case, $(N(c)-c) \cap R$ is diffeomorphic to $c' \times (0,1)$ where $c'$ double covers $c$. We choose $v_1$ and $v_2$ in $c'$ which cover $v_c$ and let $\underline{v}_i$ and $\underline{c}_i$ be the diffeomorphic images of $v_c \times \varepsilon$ and the two arcs of $c' \times \varepsilon - \{v_1 \times \varepsilon, v_2 \times \varepsilon\}$. The four cases are:

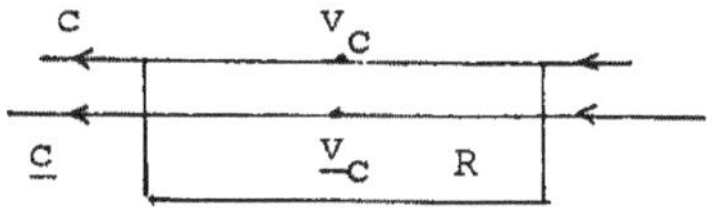


$(i_c)$

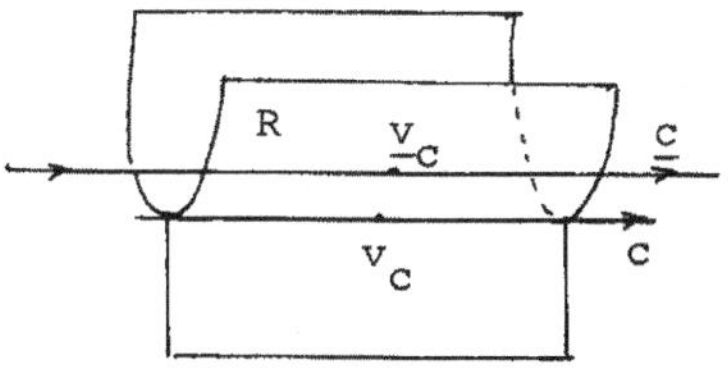


$(ii_c)$

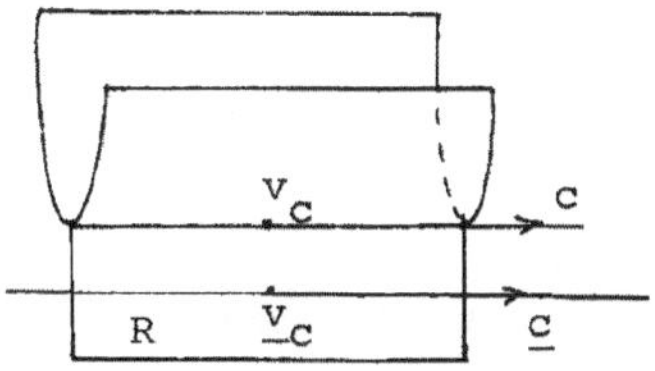


$(iii_c)$

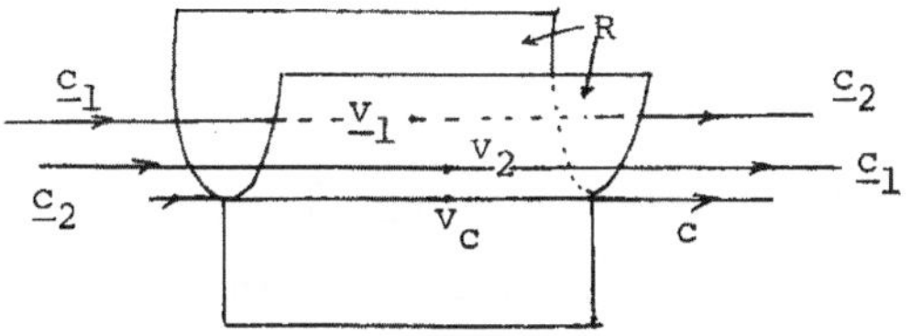


$(iv_c)$

In the first three cases the lifts have the form $\hat{c}\hat{v}_c$ and in the last the lift has the form $\hat{c}_1\hat{v}_1\hat{c}_2\hat{v}_2$. The figures in the transverse manifolds are respectively:

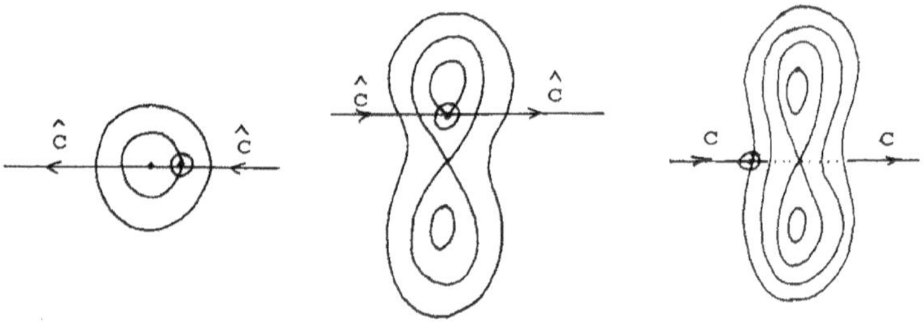


$(i_c)$ $(ii_c)$ $(iii_c)$

(If $B(c)$ is trivial)

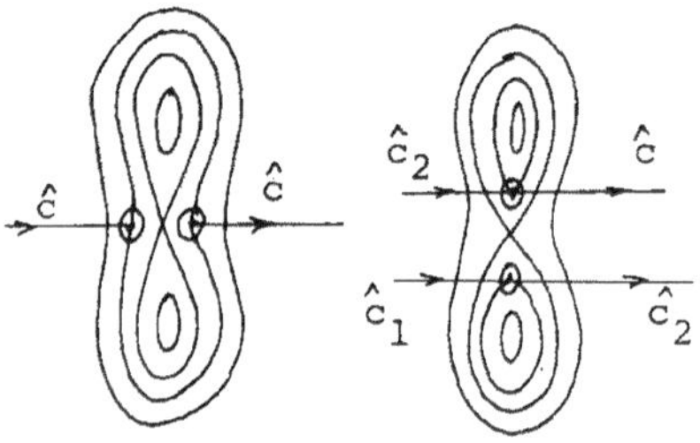


$(iii_c)$ $(iv_c)$

(If $B(c)$ is non-trivial)

Thus in these cases $w_{v_c}(\hat{v})$ are:

$(i_c)$ $u^{\pm k}$, where $u$ is the symbol for the whole fibre $q^{-1}(\underline{v}_c)$

$(ii_c)$ $a^{\pm k}$, where $a$ is the symbol for the whole fibre $q^{-1}(\underline{v}_c)$

$(iii_c)$ $a(ba)^{\pm k}$, where $(ba)$ is $w_{v_c}(R)$

$(iv_c)$ $w_{v_c}(\hat{v}_1)$ is $a^{\pm k}$, $w_{v_c}(\hat{v}_2)$ is $b^{\pm k}$.

Now let $b_i \in H_1(R)$ be a generator which represents the class of oriented curve $\beta_i$ in $\partial R$. Given $(f,h) \in I_S$, we evaluate $(P_{Th} \circ \phi_R)_*(b_i) = s_h(R)(b_i)$. This value is the same as the degree of the map

$$P_{Th} \circ \phi_R \circ \underline{\beta}_i : S^1 \longrightarrow S^1.$$

However each of the $\phi_R \circ \underline{\beta}_i$ is smoothly homotopic to one of our lifts, each of which has the form $\hat{c}_0\hat{v}_0 \ldots \hat{c}_k\hat{v}_k$. Thus we must examine the degree of $P_{Th} \circ (\hat{c}_0 \ldots \hat{v}_k)$.

<u>Definition</u>. If $g : [t_0,t_1] \longrightarrow S^1 : t \longrightarrow e^{2\pi i\gamma(t)}$, we let

$$\alpha(g) = \gamma(t_0) - \gamma(t_1).$$

Using this notation,

$$s_h(R)(b) = \sum_{j=0}^{k} \alpha(P_{Th}\circ\hat{c}_h) + \sum_{j=0}^{k} \alpha(P_{Th}\circ\hat{v}_j).$$

We interpret each of these sums: By our construction of the lifts $\hat{c}_j$, $\sum_{j=0}^{k} \alpha(P_{Th}\circ\hat{c}_j)$, has two parts: the first is

$$\sum n(R,c)\tau_h(c),$$

where the sums extend over all elements $c \in C$ in the $R$ represented by $b$, the second part is a sum of $(\pm 1/4)$'s, one for each cusp that lies on the boundary component as in figure (ix). Each such cusp, $v$ contributes $\pm\frac{1}{4}\delta_h(v)$, the sign depending only on the lift--not on $h$. Thus

$$\sum_j \alpha(P_{Th}\circ\hat{c}_j) = \sum n(R,c)\tau_h(c) + \sum \varepsilon_v\delta_h(v).$$

To interpret $\alpha(P_{Th}\circ\hat{v}_j)$, there are two cases to consider.

(a) If $v_j$ is a cusp lying on the boundary component as in figure (viii), $\hat{v}_j$ is the closed curve traversing the oriented fibre $q^{-1}(\underline{v}_j)$, $k(v_j,R)$ times. In that case

$$\alpha(P_{Th}\circ\hat{v}_j) = k(v_j,R)\delta_h(v_j).$$

(b) Otherwise, $w_{v_j}(\hat{v}_j)$ has form $a_\ell\ldots a_m(a_0\ldots a_m)^{k(v_j,R)}$ where $a_0\ldots a_m$ is the word in the arclabels for $A_{v_j}$ which represents $R$. Here $a_\ell\ldots a_m$ is a terminal segment of $a_0\ldots a_m$. The curve $\hat{v}_j$ is denoted by $\langle w_{v_j}(\hat{v}_j)\rangle$.

$$\alpha(P_{Th}\circ\langle a_\ell\ldots a_m(a_0\ldots a_m)^{k(v_j,R)}\rangle$$

$$= k(v_j,R)r_h(R) + \sigma_h[a_\ell\ldots a_m] + \tfrac{1}{2}(u_{\delta_h}(a_{\ell-1},a_\ell) + u_{\delta_h}(a_m,a_0))$$

(See §3.6 and 3.7 for the definitions of $\sigma_h$ and $u_{\delta_h}$ respectively.)

Combining this description of the value of $s_h(R)(b_i)$ with the results of §3.7, we see that for each $R \in \mathcal{R}$ and $i = 1,\ldots,n(R)$ we have found linear functions in the values of $\delta$ and $\nu$, say $E_j(R)$ such that $E_j(R)(\delta_h,\nu_h) = s_h(R)(b_j)$. The functions $E_j(R)$ depend only on the lift and deformation of the arcs of $\partial R$ and not on $h$. To conform to our original notation we define $\tilde{s} : L'_\Sigma \longrightarrow H^1(\mathcal{R},\mathbb{Z})$ by setting:

$$\tilde{s}(\delta,\nu)(R)(d_j) = E_j(R)(\delta,\nu), \quad j = 1,\ldots,n(R).$$

Remarks.

1) Since $N^*_\delta \subseteq N'_\delta$ is given as the set of zeros of linear map in the values of $\delta$ and $\nu$, we see that $N^*_\delta$ is the translation of a submodule of $(C \cup A^*, \mathbb{Z})$.

2) By considering the two consistency conditions defining $N'_\delta$ and the linear condition in $\delta$ and $\nu$ defining $N^*_\delta$ we see that $N^*_{-\delta} = -N^*_\delta$.

## 3.9 The last invariant of $[I_\Sigma]_\Sigma$; $[I_\Sigma]_\Sigma \longleftrightarrow Q$

In this section we classify the fibre of the surjective map of the corollary of §3.8.

$$\pi = (\lambda,\zeta) : [I_\Sigma]_\Sigma \longrightarrow (L\times Z) = P.$$

For each $p \in P$, let $(f,h_p) \in I_\Sigma$ be chosen so that $\pi(f,h_p) = p$. Let $h$ $[h_p]$. By definition of $I_\Sigma$, we know that $(h)_{\hat{\Sigma}} = (h_p)_{\hat{\Sigma}}$, so we may assume that $h$ and $h_p$ agree on a neighborhood $B(\Sigma)$ of $\hat{\Sigma}$. Let $(M-B(\Sigma))^2$ be the double of $M - B(\Sigma)$ and define the map

$$h \cdot h_p : (M - B(\Sigma))^2 \longrightarrow \mathbb{R}^2$$

which maps one copy of $M - B(\Sigma)$ by $h$ and the other by $h_p$. $M - B(\Sigma)$ is the union of trivial circle bundles $B(R)$ over surfaces, $\tilde{R} \subseteq R$ for all $R \in \mathcal{R}$ where $\tilde{R} = R - N(\Sigma)$ is homeomorphic to $R$ but with smooth boundary. Thus $(M - B(\Sigma))^2 = \bigcup_{R\in\mathcal{R}} (B(\tilde{R}))^2$. Each component $(B(\tilde{R}))^2$ is a circle bundle over the double $\tilde{R}$, and the map $(h\cdot h_p)$ restricts to an immersion on each fibre circle of $(B(R))^2$. Thus we can define:

$$P_{T(h\cdot h_p)} : (M - B(\Sigma))^2 = \bigcup_{R\in \mathcal{R}} (B(\tilde{R}))^2 \longrightarrow S^1$$

as in §3.1.

Letting $\phi_R : R \times S^1 \longrightarrow B(R)$ be our usual trivialization we have:

$$(P_{T(h\cdot h_p)} \circ \phi_{\tilde{R}^2}) : \tilde{R}^2 \times S^1 \longrightarrow S^1.$$

Passing to the induced map on homology we obtain:

$$\begin{array}{ccccc} & & H_1(S^1) & & \\ & & \uparrow & \searrow^{r_{(h\cdot h_p)}(R)} & \\ (P_{T(h\cdot h_p)} \circ \phi_{\tilde{R}^2})_* & : & H_1(\tilde{R}^2) \oplus H_1(S^1) & \longrightarrow & H_1(S^1) \\ & & \downarrow & \nearrow_{s_{(h\cdot h_p)}(R)} & \\ & & H_1(\tilde{R}^2) & & \end{array}$$

The map $r_{(h\cdot h_p)}(R)$ is just the common value $r_h(R) = r_{h_p}(R)$. However the map $s_{(h\cdot h_p)}(R)$ captures something new. From the Mayer-Vietoris sequence, the values of $s_{(h\cdot h_p)}(R)$ on the image of $H_1(\tilde{R}) \oplus H_1(\tilde{R})$ in $H_1(\tilde{R}^2)$ are determined by $s_h(R) = s_{h_p}(R)$. However the new information comes from the value of $s_{(h\cdot h_p)}(R)$ on $H_1(\tilde{R}^2)/\mathrm{Im}(H_1(\tilde{R}) \oplus H_1(\tilde{R}))$. We describe a set of generators for this quotient. A simple arc joining any pair of components of $\partial\tilde{R}$ determines an element of $H_1(\tilde{R}^2)$. Label the components of $\partial\tilde{R}$ as $\beta_1,\ldots,\beta_{n(R)}$ and the elements of $H_1(R^2)$ determined by an arc joining $\beta_i$ to $\beta_{i+1}$ as $\alpha_i$, $i = 1, \ldots, n(R)-1$. The set $\{\alpha_i, i = 1,\ldots,n(R)-1\}$ generates $H_1(\tilde{R}^2)/\mathrm{Im}(H_1(R) \oplus H_1(R))$.

<u>Definition</u>. Let $(f,h) \in I_\Sigma$ and $p = \pi(f,h)$. Define:

$$\gamma : I_\Sigma \longrightarrow \times_{R\in\mathcal{R}} (\mathbb{Z}^{n(R)-1}) \quad \text{by}$$

$$\gamma(f,h) = (s_{h\cdot h_p}(R)(\alpha_1),\ldots,s_{h\cdot h_p}(R)(\alpha_{n(R)-1})),$$

where $n(R)$ is the number of boundary components of $R$ and $\{\alpha_i\}$ are the elements of $H_1(\tilde{R}^2)$ determined by the arcs joining the boundary components.

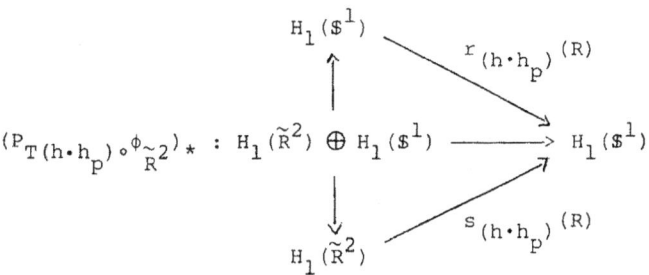

<u>Notation</u>. We let $G = \bigcup_{R\in\mathcal{R}} (\mathbb{Z}^{n(R)-1})$ and $\mathcal{Q} = \mathcal{P} \times G = (L\times Z\times G)$.

To see that $\gamma$ is well defined we check the following: Suppose we chose another neighborhood $B'(\Sigma) \subseteq B(\Sigma)$ so that $B(\Sigma) - B'(\Sigma)$ is a collar neighborhood of $\partial B(\Sigma)$. Let $(h\cdot h_p)'$ be the map constructed from $h$, $h_p$ restricted to $M - B'(\Sigma)$. We claim that $(P_{T(h\cdot h_p)})_* = (P_{T(h\cdot h_p)'})_*$. It is enough to check that $s_{(h\cdot h_p)}(R)(\alpha_i) = s_{(h\cdot h_p)'}(R)(\alpha_i)$. Let $M - B'(\Sigma) = \bigcup_{R\in\mathcal{R}} B(R')$. We know that $R' - R$ is a collar neighborhood of $\partial R$; thus we write $R' = R \cup (\partial R\times I)$, where $R$ is identified with $\partial R \times 0$ and $\partial R'$ with $\partial R \times 1$. Suppose $a_i$ is any arc in $R$ joining boundary components $\beta_i$ to $\beta_{i+1}$. An arc $a_i'$ joining the corresponding components of $\partial R'$, $\beta_1$ and $\beta_{i+1}$ may be constructed by preceding the arc $a_i$ by an arc $t \longrightarrow (x_i,(1-t)) \in \partial R \times I$, where the initial point of $a_i$ is $x_i \in \beta_i$ and following $a_i$ by an arc $t \longrightarrow (x_{i+1},t)$ where $x_{i+1} \in \beta_{i+1}$ is the terminal point of $a_i$. In computing $s_{(h\cdot h_p)'}(R)(\alpha_i)$ we must compute the total turning of the $T(h\cdot h_p)'$ image of the unit tangent vector to the q-fibre above the double of $a_i'$. However $h = h_p$ on $R' - R$, hence the contribution of $S_{(h\cdot h_p)}(R)(a_i')$ that comes from the arc $(x_i,(1-t))$ and $(x_{i+1},t)$ are zero since in the double of $a_i'$ we traverse each of these arcs twice in opposite directions. Thus $\gamma$ is well defined. As an obvious corollary to this argument we see that for any $p \in \mathcal{P}$, $\gamma(f,h_p) = 0$.

<u>Theorem</u>. $\pi \times \gamma : I_\Sigma \longrightarrow \mathcal{Q}$ <u>is constant on classes in</u> $[I_\Sigma]_\Sigma$ <u>and defines a 1:1 correspondence</u>

$$(\pi\times\gamma) : [I_\Sigma]_\Sigma \longrightarrow \mathcal{Q}.$$

<u>Proof</u>. Suppose $[f,h]_\Sigma = [f,h']_\Sigma \in [I_\Sigma]_\Sigma$. Then $\pi(f,h) = \pi(f,h')$. Call the common value, $p$. We know that $h$ and $h'$ are f-fibre regularly homotopic where the homotopy $H : h \simeq h'$ is fixed on a neighborhood of $\hat{\Sigma}$, $B(\Sigma)$. Thus the maps, $h\cdot h_p$ and $h'\cdot h_p$ are homotopic via $H_t\cdot h_p$. But since $(P_{T(H_t\cdot h_p)})_*$ is independent of $t$, $\gamma(f,h) = \gamma(f,h')$. So $(\pi\times\gamma) : [I_\Sigma]_\Sigma \longrightarrow \mathcal{Q}$ is well defined.

(i) $\pi \times \gamma$ <u>is surjective</u>. Let $(p,g) \in \mathcal{Q}$. By modifying $h_p$ on a neighborhood of the boundary of a neighborhood $B(\Sigma)$ of $\hat{\Sigma}$, we construct a map $h_{(p,g)}$ such that $(\pi,\gamma)(f,h_{(p,g)}) = (p,g)$. For each $R \in \mathcal{R}$, the boundary components $\beta_1,\dots,\beta_n(R)$ of $\tilde{R}$ are boundary components of $B(\Sigma)$. Take a little neighborhood of $B(\beta_i) = q^{-1}(\beta_i)$ diffeomorphic to $B(\beta_i) \times I$. We suppress the diffeomorphism of $B(\beta_i) \times I$ into $M$, so we have $h_p : B(\beta_i) \times I \longrightarrow \mathbb{R}^2$ which we change as follows. Suppose $g(R) = (g_1,\dots,g_{n(R)-1}) \in \mathbb{Z}^{n(R)-1}$, and define $k_i = \sum_{j=1}^{i-1} g_j$, $i = 2,\dots,n(R)$ and $k_1 = 0$. That is, $g_i = k_{i+1} - k_i$, $i = 1,\dots,n(R)-1$, and $k_1 = 0$. Let $\phi : [0,1] \circlearrowleft$ be a smooth, non-decreasing function that vanishes in a neighborhood of $0$ and is $1$ in a neighborhood of $1$. We define $h_{(p,g)} : B(\beta_j) \times I \longrightarrow \mathbb{R}^2$ : $(x,t) \longrightarrow \rho_j(h_p(x,t),t)$, where $\rho_j : (\mathbb{R}^2 = \mathbb{C}) \times I \longrightarrow (\mathbb{R}^2 = \mathbb{C})$ : $(z,t) \longrightarrow e^{2\ ik_j(t)}z$, and elsewhere let $h_{(p,g)} = h_p$. On a neighborhood of $\hat{\Sigma}$ and on the complement of a slightly larger neighborhood the $\hat{\Sigma}$, $h_{(p,g)} = h_p$. In the difference of the two neighborhoods the "g-turns" are inserted. Obviously $(\pi\times\gamma)(f,h_{(p,g)}) = (p,g)$.

// ($\pi \times \gamma$ surjective)

(ii) $(\pi\times\gamma)$ <u>is injective</u>. If $(\pi\times\gamma)(f,h) = (p,g)$ we must show that $[f,h]_\Sigma = [f,h_{(p,g)}]_\Sigma \in [I_\Sigma]_\Sigma$. As in the construction of $h_{(p,g)}$ we write $M$ as the union of $B(\Sigma)$, $\cup\, B(\tilde{R})$, $\cup\, (B(\partial\tilde{R})\times I)$ where $B(\partial\tilde{R}) = q^{-1}(\partial\tilde{R})$ is identified with $B(\partial\tilde{R}) \times \{1\}$ and $\partial B(\Sigma) = \cup$ $B(\partial R) \times \{0\}$. (We suppress the embedding diffeomorphisms of $B(\partial R) \times I$ in $M - B(\Sigma)$.) We may assume that $h = h_p = h_{(p,g)}$ on a neighborhood of $\overline{B(\Sigma)}$, and $h_p = h_{(p,g)}$ in $\cup\, B(\tilde{R})$. We construct an $f$-fibre regular homotopy $H : h \simeq h'$ where $h = h'$ on $B(\Sigma)$ and $h' = h_p$ on $\cup\, B(\tilde{R})$. Since $(f,h) = p$, we know that there is an $f$-fibre regular homotopy $K : (M-B(\Sigma)) \times I \longrightarrow \mathbb{R}^2$ connecting $h \mid M - B(\Sigma)$ to $h_p \mid M - B(\Sigma)$. Define $H : M \times I \longrightarrow \mathbb{R}^2$ by $H \mid B(\Sigma) \times I = h \mid B(\Sigma)$, $H \mid B(\tilde{R}) \times I = K \mid B(\tilde{R}) \times I$. To define $H$ on $(B(\partial\tilde{R})\times I)$ we need two smoothing functions $\psi_0 : I \longrightarrow I$ and $\psi_1 : I \longrightarrow I$, both smooth, non-decreasing functions such that $\psi_0 \mid [0,\varepsilon] = \psi_1 \mid [1-\varepsilon,1] = 0$ and $\psi_0[\varepsilon',1] = \psi_1[0,1-\varepsilon'] = 1$ where $0 < \varepsilon < \varepsilon' < 1/2$. For $(x,s) \in B(\partial\tilde{R}) \times I$,

$$H((x,s),t) =$$

$$\tfrac{1}{2}\{(1-\psi_0(s))h(x,s) + (\psi_0(s) + \psi_1(s))K(x,s,\psi_0(s)t) + (1-\psi_1(s))K(x,s,t)\}.$$

By choosing $\varepsilon'$ small enough this obviously defines an $f$-fibre regular homotopy between $h$ and some element $h' \in F$. Since nothing changes on $B(\Sigma)$, we see $h = h'$ there. On $B(R)$, $h' = h_p$.

Furthermore on a little neighborhood of $\bigcup_R B(\partial\tilde{R}) \times \{0\} = \partial B(\Sigma)$, $h' = h = h_p$ and on a little neighborhood of $\cup(B(\partial\tilde{R}) \times \{1\} = \partial(\cup B(\tilde{R}))$, $h' = h_p$. Thus we have shown that we may assume that $h = h_p$ on a neighborhood of $(\overline{B(\Sigma)} \cup \cup \overline{B(\tilde{R})})$. Without changing the class of $(f,h)$, we normalize $h$ a little more. We consider the map

$$P_{T(h\cdot h_p)} : (B(\partial\tilde{R})\times I)^2 \longrightarrow \$^1.$$

This makes sense since $h$ and $h_p$ agree on a neighborhood of $B(\partial\tilde{R})$ $\{0,1\}$. Since $(B(\partial\tilde{R})\times I)^2 = B(\partial R) \times \$^1$ we see that the degree of this map restricted to $z_j \times \$^1$ gives an integer $k_j$ for each component of $B(\partial\tilde{R})$ for $z_j \in B(\beta_j)$ for each component, $\beta_j$ of $\partial\tilde{R}$. Claim: $g_j = k_{j+1} - k_j$, $j = 1,\dots,n(R)$. Let $\beta_j$ and $\beta_{j+1}$ be components of $\partial(M-B(\Sigma))$. Let $\alpha_j$ be an arc joining them. We may take the path $\alpha_j$ to be of the form: $((x_j,1),3t)$ where $(x_j,1) \in \beta_j \times \$^1 \subseteq \partial\tilde{R} \times \$^1 \cong B(\partial\tilde{R})$ for $0 \leqslant t \leqslant 1/3$, for $t \in [1/3,2/3]$, $\alpha_j$ is a path in $B(\tilde{R})$ joining $((x_{j,1}))$ with $(x_{j+1},1) \in \beta_{j+1} \times S^1 \subseteq (\partial\tilde{R}\times\$^1)$, and the final segment for $t \in [2/3,1]$, $((x_{j+1},1),3(1-t))$. We know since $\gamma(f,h) = g$, that the degree of

$$P_{T(h\cdot h_p)} : (B(\tilde{R} \cup (\partial\tilde{R}\times I))^2 \longrightarrow \$^1$$

restricted to the double of $\alpha_j$ is $g_j$. However since $h = h_p$ on $B(\tilde{R})$, we see that the only contribution to this degree that doesn't cancel is the length of the arc swept out by the image under $P_{T(h\cdot h_p)}$ of the double of the last third of $\alpha_j$ minus that of the first third of $\alpha_j$, which is precisely $k_{j+1} - k_j$.

Note: Let $\mu : I \longrightarrow B(\partial\tilde{R}) \times I : t \longrightarrow ((x,1),t)$ for $x \in \partial\tilde{R}$. Let $\mu^2$ be the double of this, taking $\$^1 \longrightarrow B(\tilde{R}) \times \$^1$. To compute the degree of $P_{T(h\cdot h_p)}\circ\mu^2 : \$^1 \longrightarrow \$^1$ we compute the length of the arc swept out by $P_{Th}\circ\mu : I \longrightarrow \$^1$ and subtract from it the length of arc swept out by $P_{Th_p}\circ\mu : I \longrightarrow \$^1$ which is the same as joining two copies of $I$ together at $0$ and mapping the first half of the resulting interval by $P_{Th_p}\circ\mu$ and the second half by $P_{Th}\circ\mu$ and taking the length of arc swept out by this map.

Our object now is to construct an $f$-fibre regular homotopy of $h$ so that the integer just computed will be altered so that $k_1 = 0$. Let $\phi : [0,1] \circlearrowleft$ vanish on $[0,\varepsilon]$ and be $1$ on $[1-\varepsilon,1]$, and be smooth and non-decreasing. Define $H : h \simeq h'$ by

$$\begin{cases} (x,t) \longrightarrow h(x), & \text{for} \quad (x,t) \in B(\Sigma) \times J \\ ((y,s),t) \longrightarrow e^{-2\pi i k_1 t\phi(s)} h(y,s), & \text{for} \quad ((y,s),t) \in (B(\partial R)\times I) \times J \\ (x,t) \longrightarrow e^{-2\pi i k_1 t} h(x), & \text{for} \quad (x,t) \in \cup B(R) \times J \end{cases}$$

If we let $k_j'$ be the degrees computed using the map $h'$ analogous to $k_j$ for $h$ it is obvious that $k_j' = k_j - k_1$. Thus it is no restriction to assume that the integers computed for $h$ satisfy: $k_{j+1} - k_j = g_j$, and $k_1 = 0$.

For such an $h$ we already knew that $h = h_p = h(p,g)$ on $B(\Sigma) \cup (\cup B(R))$. With this normalization of the $k_j$ we also know that the map $P_{T(h \cdot h_{(p,g)})} : B(\partial\tilde{R}) \times \$^1 \longrightarrow \$^1$ had degree zero on each $x \in \$^1$ for $x \in B(\partial\tilde{R})$. Thus we have two maps $h, h_{(p,g)} : B(\partial\tilde{R}) \times I \longrightarrow \mathbb{R}^2$ which agree on a neighborhood of $(B(\partial\tilde{R})\times\partial I)$. It suffices to construct an f-fibre regular homotopy between $h$ and $h_{(p,g)}$ on $B(\partial\tilde{R}) \times I$ which is constantly $h$ on a neighborhood of $(B(\partial\tilde{R})\times\partial I)$.

The proof of the existence of such a homotopy is a straightforward application of the generalized Whitney-Graustein Theorem in §3.6 and proven in the appendix. // $((\pi\times\gamma)$ is injective)

## 3.10 The fibre of $[I_\Sigma]_\Sigma \longrightarrow [I]$

We summarize what we have shown so far about the various equivalences of immersions over f.

$$\begin{array}{ccccccc}
[I_V]_V & \xleftrightarrow{3.5} & [I_S]_S & \xleftrightarrow{3.6} & [I_\Sigma]_\Sigma & \xleftrightarrow[(\pi,\gamma)]{3.9} & \mathcal{Q} \\
{\scriptstyle P_V}\downarrow & & & & \Big\downarrow & & \\
[I_V]_E & \xleftrightarrow{3.4} & [I_E]_E & & \Big\downarrow {\scriptstyle P} & & \\
{\scriptstyle P_E}\downarrow & & & & \Big\downarrow & & \\
[I_V] & \xleftrightarrow{3.4} & [I_E] & \xleftrightarrow{3.3} & [I] & &
\end{array}$$

The map $P$ makes the diagram commute, the other vertical maps are the obvious surjective maps obtained by weakening equivalence relations.

We can introduce an equivalence relation in $\mathcal{Q}$ by declaring $q_1 \sim q_2$ iff $P\circ(\pi,\gamma)^{-1}(q_1) = P\circ(\pi,\gamma)^{-1}(q_2)$. Our object in this section and the next is to describe the fibre of $P$ and to describe the equivalence relation in $\mathcal{Q}$ explicitly. Since studying P is equivalent to studying $P_E \circ P_V$ we break the study of the fibre into two parts --of $P_E$ and of $P_V$.

N.B. In this section $E : S(f) \longrightarrow \mathbb{R}^4$ will be the immersion described at the end of §3.3 and denoted there as $E_i$. Everything that has been done until this point is valid for any choice of immersion of S(f) over f. The coordinate expressions used in §3.4 in the definition of $I_{V_*}$ and in §3.5 would have to be changed as have been indicated there.

To see that the surjective map $P_V : [I_V]_V \longrightarrow [I_V]_E$ is not injective note that for $(f,h) \in I_V$, $\tau_h$ is constant on all elements in $[f,h]_V$ but is not constant on the elements in $[f,h]_E \in [I_V]_E$. To alter h so that we change the $[I_V]_V$-class but not the $[I_V]_E$ class, just spin the h-target-$\mathbb{R}^2$-space at any non-cusp point in $\hat{V} = q^{-1}(V) \cap S(f)$. To alter its $[I_V]_E$ requires spinning at a cusp point. We will make such "spinning" precise and show that it accounts for the whole fibre of P.

Let $(f,h_0)$ and $(f,h_1) \in I_V$ and suppose $[f,h_0] = [f,h_1] \in [I_V]$. Let H be an f-fibre regular homotopy between $h_0$ and $h_1$. For any point $x \in \hat{V}$ and any non-zero vector $\xi \in \ker Tf_x$, $TH_.(\xi) : I \longrightarrow \mathbb{R}^2 - \{0\} : t \longrightarrow TH_t(\xi)$. Since $(f,h_0)$ and $(f,h_1) \in I_V$, we know that $TH_0(\xi) = TH_1(\xi)$, hence $TH_.(\xi)/|TH_.(\xi)|$ defines a map from $S^1$ to $S^1$ whose degree is independent of the choice of $\xi \in \ker Tf_x$. This defines an integer valued function which we denote by:

$$sp_H : \hat{V} \longrightarrow \mathbb{Z}.$$

Remarks.

1. If $[f,h_0]_E = [f,h_0]_E$ and we had chosen $H : h_0 \underset{E}{\simeq} h_1$ (i.e. $(f,H_t) \in I_E$ for all $t \in [0,1]$) then if $x = q^{-1}(v) \cap S(f)$ and v is a cusp then $sp_H(x) = 0$. This is clear since the tangent line to S(f) at x is in $\ker Tf_x$ and is fixed under $TH_t$ since $(f,H_t) \in I_E$.

2. Let $H^0 : h_0 \simeq h_1$ and $H^1 : h_1 \simeq h_2$. If $H^1 * H^0$ is the usual concatenation of homotopies, then

$$sp_{H^1 * H^0} = sp_{H^1} + sp_{H^0}.$$

Proposition 0. If $[f,h_0]_E = [f,h_1]_E \in [I_V]_E$ then there is an $H : h_0 \underset{E}{\simeq} h_1$ with $sp_H(z) = 0$, for some $z \in \hat{V}$. Equivalently, if there are no cusps, given $\tilde{H} : h_0 \underset{E}{\simeq} h_1$, $k \in \mathbb{Z}$, $\exists\ H : h_0 \underset{E}{\simeq} h_1$ such that $sp_H = sp_{\tilde{H}} + k$.

Proof. As already remarked if there are any cusps, then for any such $H : h_0 \underset{E}{\simeq} h_1$, $sp_H(z) = 0$ for $z \in q^{-1}(C) \cap S(f) = \hat{C}$. On the other hand if there are no cusps then our immersion $E \mid S(f) = (f,0) \mid S(f)$ and so if $\tilde{H} : h_0 \underset{E}{\simeq} h_1$ and $sp_H(z) = k$ for some $z \in V$, then

$$H_t = {}^{-2\pi ikt}\tilde{H}_t. \quad //$$

N.B. It is this proposition and proof that motivated replacing the embedding of $S(f)$ by the immersion (which we still call $E$). I think the proposition is true with $E$ an embedding but I have not translated the pictorial idea of a proof into a proof.

Proposition 1. If $(f,h) \in I_V$ and $H : h \simeq h$ is an f-fibre regular homotopy then $sp_H$ is constant.

Proof. Since $H$ connects $h$ with itself, we see that for any $x \in M$ and any non-zero element $\xi \in \ker Tf_x$, $TH_{\cdot}(\xi) : I \longrightarrow \mathbb{R}^2 - \{0\}$ defines a map from $S^1$ to $S^1$ and hence we have an integer valued continuous function defined on $\bigcup_{x \in M} (\ker Tf_x - \{0\})$. Since this set is connected, this function is constant. //

Definition. Two functions $\alpha, \beta : \hat{V} \longrightarrow \mathbb{Z}$ are equivalent if $\alpha - \beta$ is constant. Denote the equivalence class of $\alpha$ by $[\alpha]$.

Corollary 1. If $(f,h_i) \in I_V$, $i = 0, 1$ and $H$ and $H' : h_0 \simeq h_1$ are f-fibre regular homotopies then

$$[sp_H] = [sp_{H'}].$$

Proof. Let $K : h_1 \simeq h_0$ be defined by $K_t = H'_{1-t}$ then $K*H : h_0 \simeq h_0$ so

$$(\text{constant}) = sp_{K*H} = sp_K + sp_H = -sp_{H'} + sp_H. \quad //$$

Thus we have shown that $[sp_H]$ for $H : h_0 \simeq h_1$ an f-fibre regular homotopy between $h_0$ and $h_1$ where $(f,h_i) \in I_V$ depends only on $(h_0,h_1)$ and not on $H$. Thus we write $[sp_{(h_0,h_1)}]$ for any $[f,h_0] = [f,h_1] \in I_V$.

Theorem 1. Let $(f,h_i) \in I_V$, $i = 0, 1, 2$

1) If $[f,h_0] = [f,h_1] = [f,h_2] \in I_V$, then $[f,h_1]_V = [f,h_2]_V$ iff $[sp_{(h_0,h_1)}] = [sp_{(h_0,h_2)}]$.

2) *If* $(f,h) \in I_V$ *and* $n : \hat{V} \longrightarrow \mathbb{Z}$ *is given then there is an* $(f,h') \in I_V$ *with* $[f,h] = [f,h'] \in [I_V]$ *and* $[sp_{(h,h')}] = [n]$.

An alternative statement of this theorem is:
*If* $P : [I_V]_V \longrightarrow [I_V] : [f,h]_V \longrightarrow [f,h]$, *then for any* $(f,h) \in I_V$, *the map*

$$P^{-1}([f,h]) \longrightarrow [\hat{V},\mathbb{Z}] : [f,h']_V \longrightarrow [sp_{(h,h')}]$$

*is a 1:1 correspondence.*

We break the theorem into two parts, one corresponding to $P_V : [I_V]_V \longrightarrow [I_V]_E$ and the other corresponding to $P_E : [I_V]_E \longrightarrow [I_V]$. Let $\hat{C} = q^{-1}(C) \cap S(f)$, where $C$ is the set of cusps of $f$ in $W$.

If $(f,h)$ and $(f,h') \in I_V$ and $[f,h]_E = [f,h']_E$, then by $\{sp_{(h,h')}\}_E$ we mean the set of all functions $sp_H : \hat{V} \longrightarrow \mathbb{Z}$ where $H : h \underset{E}{\simeq} h'$. We know that all of the functions in $\{sp_{(h,h')}\}_E$ are equivalent, but if $C \neq \emptyset$ we know $\{sp_{(h,h')}\}_E$ is a singleton since any such function must vanish on $\hat{C}$.

*Theorem 2.* *Let* $(f,h) \in I_V$. *The following maps are 1:1 correspondences.*

$(V_0)$ *If* $C = \emptyset$, *then*

$$P_V^{-1}[f,h]_E \longrightarrow [\hat{V},\mathbb{Z}]$$
$$[f,h']_V \longrightarrow [\{sp_{(h,h')}\}_E]$$

$(V_1)$ *If* $C \neq \emptyset$, *then*

$$P_V^{-1}[f,h]_E \longrightarrow ((\hat{V},\hat{C}),(\mathbb{Z},0))$$
$$[f,h']_V \longrightarrow \{sp_{(h,h')}\}_E$$

(E) $$P_E^{-1}[f,h] \longrightarrow [\hat{C},\mathbb{Z}]$$
$$[f,h']_E \longrightarrow [sp_{(h,h')} \mid \hat{C}].$$

*Proof.* Theorem 2 implies Theorem 1.

*Injectivity* of $P^{-1}[f,h] \longrightarrow [\hat{V},\mathbb{Z}]$

Let $[f,h_i] = [f,h] \in [I_V]$, $i = 1, 2$ and suppose $[sp_{(h,h_1)}] = [sp_{(h,h_2)}]$, then since $[sp_{(h,h_1)} \mid \hat{C}] = [sp_{(h,h_2)} \mid \hat{C}]$, (E) gives $[f,h_1]_E = [f,h_2]_E$. Since $[sp_{(h,h_1)}] = [sp_{(h,h_2)}]$, $[sp_{(h_1,h_2)}] = [0] = [sp_{(h_1,h_1)}]$. By either version of (V), since $[f,h_2]_V \in P_V^{-1}[f,h_1]_E$ we see that $[f,h_1]_V = [f,h_2]_V$.

<u>Surjectivity</u> of $P^{-1}[f,h] \longrightarrow [\hat{V},\mathbb{Z}]$.
Choose any $g : \hat{V} \longrightarrow \mathbb{Z}$. By (E) there is an $(f,\tilde{h}) \in I_V$ such that $[f,\tilde{h}] \in P_E^{-1}[f,h]$ and $[sp_{(h,\tilde{h})} \mid \hat{C}] = [g \mid \hat{C}]$. Let $H : h \simeq h'$ be an f-fibre regular homotopy. If $\hat{C} = \emptyset$ by $(V_0)$, there is an $(f,h') \in I_V$ with $[f,h']_V \in P^{-1}[f,h]_E$ and $[\{sp_{(h,h')}\}_E] = [sp_{(h,h')}] = [g]$. If $\hat{C} \neq \emptyset$ we know that $sp_H \mid \hat{C} = g \mid \hat{C} + k$ for some constant $k$. By $(V_1)$, there is an $(f,h') \in I_V$ such that $[f,h']_V \in P_V^{-1}[f,\tilde{h}]_E$ and an $H' : \tilde{h} \underset{E}{\simeq} h'$ such that $sp_{H'} = g + k - sp_H$. Thus $H*H' : h \simeq h'$ and $sp_{H*H'} = g + k$. That is $[sp_{(h,h')}] = [g]$. // 2 implies 1

<u>Proof of Theorem 2</u>. We prove all of the surjectivity parts of the theorem at once. We are given $(f,h) \in I_V$. We will work in a $C^2$PN of a point $p \in \hat{V}$, $(\phi,\psi,g)$. Let $k = h\circ\phi : I \times T \longrightarrow \mathbb{R}^2$, where, as usual in this context, $I = (-1,1)$ and $T$ is the transverse manifold to $q(p)$. We construct an f-fibre regular homotopy $H : h \cong h'$ such that $(f,h') \in I_V$, $sp_H(p) = n$ and $sp_H(p') = 0$ for $p' \in \hat{V} - \{p\}$. If $p \notin \hat{C}$, then $(f,H_t) \in I_E$ for all $t$. We modify $k$ on $I \times \mathbb{D} \subseteq I \times T$, where we may assume that $(0,0) = \phi^{-1}(p) \in I \times \mathbb{D}$. Since $(f,h) \in I_V$ we know that $k(u,x,y) = (x,\delta y)$ for $(u,x,y) \in I \times \mathbb{D}$ and $\delta = \pm 1$ (see §3.4). Let $\vartheta : I \longrightarrow [0,1]$ be a smooth function, vanishing on a neighborhood of $\partial I$ and $1$ on a neighborhood of $0 \in I$. Let $\rho : \mathbb{R} \longrightarrow [0,1]$, $\rho \mid (-\infty,1/4] = 1$, $\rho \mid [3/4,\infty) = 0$ and $-2 < \rho' \leqslant 0$. Define:

$$K_t(u,(x,y)) = e^{2\pi i n\vartheta(u^2)\rho(x^2+y^2/r^2)t}k(u,(x,y)),$$

where we identify $\mathbb{R}^2$ with $\mathbb{C}$ as usual and $r$ is any number between $0$ and the radius of $\mathbb{D}$. Obviously $K_t$ on a neighborhood of $\partial(I\times\mathbb{D})$ coincides with $k$ for all $t$ and $K_t(u,(0,0)) = k(u,(0,0)) = 0$. Thus we may insert $K_t\circ\phi^{-1}$ for $h$ on $k(I\times T)$ which yields a well-defined homotopy between $h$ and $h' = H_1$. The condition $K_t(u,(0,0)) = k(u,(0,0)) = 0$ says that $(f,H_t) \mid S(f) = E$ in all cases except $p \in \hat{C}$. To see that $(f,H_t) \in I$ we must check that the determinant of the Jacobian of $K_t$ with respect to $(x,y)$ is non-zero. An easy computation gives its value as $\delta$. Finally it is obvious that $sp_H(p) = n$ and $sp_H(p') = 0$ if $p' \in \hat{V} - \{p\}$. Thus we have proven surjectivity of the maps in all three cases.

We now prove:

Injectivity for $(V_0)$ and $(V_1)$.
Let $[f,h]_E = [f,h_0]_E = [f,h_1]_E$, suppose $[\{sp_{(h,h_0)}\}_E] = [\{sp_{(h,h_1)}\}_E]$. We must show that $[f,h_0]_V = [f,h_1]_V$ if there is an f-fibre regular homotopy $H : h_0 \underset{E}{\simeq} h_1$ such that $sp_H = k$. By Proposition 0, we may assume $sp_H = 0$. We will work entirely in a disc neighborhood about each $p \in \hat{V}$, $I \times \mathbb{D} \subseteq I \times T \xrightarrow{\phi} M$, $(\phi,\psi,g)$ the $C^2$PN at p. The homotopy $H : J \times M \longrightarrow \mathbb{R}^2$, $J = [0,1]$ when composed with $\phi \mid I \times \mathbb{D}$ gives a map

$$K : J \times I \times \mathbb{D} \longrightarrow \mathbb{R}^2.$$

Because of our assumption that $(f,H_t) \in I_E$ for all t we may assume that $K \mid J \times I \times 0 = 0$ if $p \notin \hat{C}$, and if $p \in \hat{C}$, then $K(t,3\varepsilon x^2,(x,0)) = (x,0)$. Also since $(f,H)$ connects elements in $I_V$ we may assume that for some $J_0$ such that $J_0 \subseteq \bar{J}_0 \subseteq J^0$, $K \mid J - J_0 \times I \times \mathbb{D}$ is independent of t and in fact if $t \in J - J_0$, that $K(t,u,(x,y)) = (x,\pm y)$. As usual it suffices to handle the case that $K(t,u,\cdot)$ is orientation preserving.

If $P \in \hat{C}$ we modify K so that $K \mid J \times I' \times 0 = 0$ on a possibly smaller interval $I' \subseteq I$ and drop the '. Expand K about 0 in $\mathbb{D}$:

$$K(t,u,x,y) = xa(t) + yb(t) + \{K(t,u,0,0) + xu\ell(t,u) + yum(t,u) + R(t,u,x,y)\}$$

where $|R(t,u,x,y)|/|(x,y)| \longrightarrow 0$ as $|(x,y)| \longrightarrow 0$. We know that for all t, $a(t) = (1,0)$ and $b(t) = b_1(t),b_2(t)$, $b_2(t) \neq 0$. Just as in the first part of the proof of the lemma of §3.4, we can replace K by

$$xa(t) + yb(t) + \vartheta(u,xa(t),yb(t))\{\quad\},$$

where $\vartheta$ vanishes in a sufficiently small neighborhood of 0 and is 1 on a neighborhood of $\partial(I\times\mathbb{D})$. Thus by shrinking the neighborhood $I \times \mathbb{D}$ as necessary, we may assume in all cases that:

a) $K \mid J \times I \times 0 = 0$,

b) $K \mid J - J_0 \times I \times \mathbb{D} = \text{proj onto } \mathbb{D}$

c) $\text{Jac} : J \longrightarrow GL(2,\mathbb{R}) : t \longrightarrow (\partial K_t(0,x,y)/\partial(x,y))_{(x,y)=(0,0)}$

is homotopic to the constant map taking J to the identify in $GL(2,\mathbb{R})$, the homotopy being fixed as the identity on $J - J_0$.

Condition c) is just the translation of $sp_H(p) = 0$. The three conditions listed are just the hypotheses of Lemma 0 of §2.1. (In the notation used there (T, J, K) are (a point, J, I) in this context. Also '$v_0 \in K$' there is '$0 \in I$' here.) The conclusions of Lemma 0 give us:

For appropriately small nested discs, $B_0 \subseteq B_1 \subseteq \mathbb{D}$, and closed nested intervals, $J_0 \subseteq J_1 \subseteq J$, $I_0 \subseteq I_1 \subseteq I$, there is an orientation preserving diffeomorphism

$$1 \times L : J \times I \times \mathbb{D} \longrightarrow \mathbb{R}^2 \times \mathbb{R}^2$$

such that $L(J\times I\times 0) = 0 \in \mathbb{R}^2$ and

1) On $J \times (I_0\times B_0)$, $L$ is projection on $B_0$

2) On $J \times I \times \mathbb{D} - J_1 \times I_1 \times B_1$, $L = K$.

From 2) we see that $L = K$ on $J \times (I\times\mathbb{D}-I_1\times B_1)$ so we define a homotopy $H'$ by replacing $H_t \mid \phi(I\times\mathbb{D})$ by $L_t\circ\phi^{-1} \mid \phi(I\times\mathbb{D})$. Also from 2), since $L = K$ on $(J-J_1) \times (I\times\mathbb{D})$ we see that the $H'$ joins $h_0$ and $h_1$. Since $L_t(I\times 0) = 0$, and 1), we know that $(f,H_t') \in I_E$ and also by 1) that the germ of $H_t'$ at $p$ is the one required there for an element of $I_V$. Thus if we alter $H$ in an analogous way at all of the points in $\hat{V}$ we obtain $(f,\tilde{H})$, $\tilde{H} : h_0 \underset{V}{\tilde{\simeq}} h_1$ as required.

// (V) injectivity

<u>Injectivity for</u> (E).

Just as in the preceding case it is enough to show that for $(f,h_i) \in I_V$ $i = 0, 1$, if $[f,h_0] = [f,h_1]$ and if for an f-fibre regular homotopy $H : h_0 \simeq h_1$, $sp_H = 0$, then we can construct $H^* : h_0 \underset{E}{\tilde{\simeq}} h_1$.

As usual it is no restriction to assume that $H_t = h_0$ for $t \in [0,\alpha)$ and $H_t = h_1$ for $t \in (1-\alpha,1]$.

<u>Lemma</u>. <u>Let</u> $(f,h_i) \in I$, $i = 0, 1$ <u>and</u> $[f,h_0] = [f,h_1]$, <u>then there is an f-fibre regular homotopy</u> $H : h_0 \simeq h_1$ <u>such that if</u> $p$ <u>is a double point or cusp point of</u> $f$, <u>then</u> $H_t(p) = h_0(p)$.

<u>Proof</u>. Let $(\phi,\psi,g)$ be a CPN such that $\phi(I\times T) \ni p$. Let $y = f(p)$, and let $U = \psi(I\times J)$ and let $\vartheta$ be a bump function on $\mathbb{R}^2$, 1 on a neighborhood of $y$ in $U$ and vanishing outside a slightly larger neighborhood of $y$ in $U$. Let $V = \phi(I\times T)$ and recall that $f(V) = U$. Let $\tilde{H} : h_0 \simeq h_1$ be an f-fibre regular homotopy, define:

$$H_t = \begin{cases} \tilde{H}_t & \text{outside of } V \\ \tilde{H}_t + (h_0(p) - \tilde{H}_t(p))\vartheta \circ f & \text{on } V \end{cases}$$

Obviously $H_t$ is smooth well defined, $H_t(p) = h_0(p)$, $H_j = \tilde{H}_j$ for $j = 0, 1$. To see that $(f, H_t) \in \mathcal{I}$ we must check that $(f, H_t)$ has maximal rank everywhere. Outside of $V$ it is automatic. On $V$ since $T(\vartheta \circ f)$ annihilates anything in $\ker Tf$, the rank $(f, H_t)$ is the same as the rank of $(f, \tilde{H}_t)$.

Thus we may assume that for any finite set of points $\{p_1, \ldots, p_k\} \subseteq S(f)$, $H_t(p_i) = h_0(p_i)$ as long as $p_i \neq p_j$ implies $q(p_i) \neq q(p_j)$. This condition is enough to guarantee that there are disjoint CPN's about the $p_i$. Thus we may assume that $H_t(p) = h_0(p)$ for all cusp points, $\hat{C}$ and all points $v^c$ for $c$ closed elements of $\mathcal{C}$, all simple double points and one of each pair of non-simple double points. That is if $p \in S(f)$ and $q^{-1}q(p) \cap S(f) = \{p, p'\}$, $p \neq p'$ we may assume that one of $H_t(p)$ or $H_t(p')$ is independent of $t$.

We finish the proof of the lemma by showing that we can find an $H$ that is constant on the remaining double points as well.

Let $p_0 \in \hat{V}$ be of type $(1 \cdot 2 \cdot 2 \cdot 1)$ or $(1 \cdot 2 \cdot 2 \cdot 3)$.

Let $\phi : I \times \mathbb{D} \longrightarrow M$ be part of a CPN for $p_0 \in S(f)$. It is no restriction to assume that $\phi(0,0) = p_0$ and that for each $u \in I$, $H_t \circ \phi \mid u \in \mathbb{D}$ is an embedding. Furthermore, the preimage of $S(f)$ is just $I \times \{0\}$, and the arc through $p_0$ is indefinite since $q^{-1}q(p_0) \cap S(f)$ has two points in it. We modify $H_t$ only on $\phi(I \times \mathbb{D})$.

Let $\mathbb{D}_0 \subseteq \mathbb{D}_1 \subseteq \mathbb{D}$ be concentric discs. We let

$$H_t' = \begin{cases} H_t & \text{outside } \phi(I \times \mathbb{D}_1) \\ H_t + (h_0(p_0) - h_t(p_0)) \circ \pi_I \circ \phi^{-1} & \text{on } \phi(I \times \mathbb{D}) \\ * & \text{on } \phi(I \times (\mathbb{D}_1 - \mathbb{D}_0)) \end{cases}$$

where $\pi_I : I \times J \longrightarrow I$ is the projection, and $\vartheta : I \longrightarrow [0,1]$ is 1 in a neighborhood of 0 and vanishes on a neighborhood of $\partial I$. We describe $*$ on $I \times (\mathbb{D}_1 - \mathbb{D}_0)$. Let $K_t = H_t \circ \phi$. We define a family of maps $\psi_t : I \times \mathbb{R}^2 \longrightarrow \mathbb{R}^2$ such that $K_t'(u,x) = \psi_t(u, K_t(u,x))$ and $H_t' \mid \phi(I \times \mathbb{D})$ is defined as $K_t' \circ \phi^{-1}$. The family of maps $\psi_t$ has the following properties:

1) $\psi_t(u, K_t(u,x)) = K_t(u,x) = (h_0(p_0) - h_t(p_0))\ (u)$ for $x \in \mathbb{D}_0$, i.e. $\psi_t(u, \cdot)$ just translates $K_t(u \times \mathbb{D}_0)$.

2) $\psi_t(u,K_t(u,x)) = K_t(u,x)$ for $(u,x)$ on a neighborhood of $\partial(I\times\mathbb{D})$.

For each $t$, $\psi_t(u,\cdot) : \mathbb{R}^2 \longrightarrow \mathbb{R}^2$ is a family of maps such that $\psi_t(u,\cdot) = 1_{\mathbb{R}^2}$ for $u$ in a neighborhood of $\partial I$. As $u$ passes from $-1$ to $1$, a singular curve appears and then disappears. This singular curve has two cusps on it. Except at the moment of creation and annihilation of the singularities, the map $\psi_t(u,\cdot)$ is stable. The introduction of the singular curve allows us to introduce a pocket in $\mathbb{R}^2$ that allows us to move $H_t(p_0)$ to $h_0(p_0)$. (For a discussion of such introduction and elimination of cusps in pairs see [$L_2$].) A $\psi_t(u,\cdot)$ in which a pocket is introduced leaves all of $\mathbb{R}^2$ fixed except $K_t(u\times\mathbb{D}_1)$ and there, maps a set containing $K_t(u\times\mathbb{D}_0)$ as suggested by the following figure where we have shaded the image of $K_t(u\times\mathbb{D}_0)$, and the heavy solid and dotted curves form the image of the singular set of $\psi_t(u,\cdot)$. It is clear that we can define such a family $\psi_t$, with which we define $H' : I \times M \longrightarrow \mathbb{R}^2$. To guarantee that the resulting $H'$ is an f-fibre

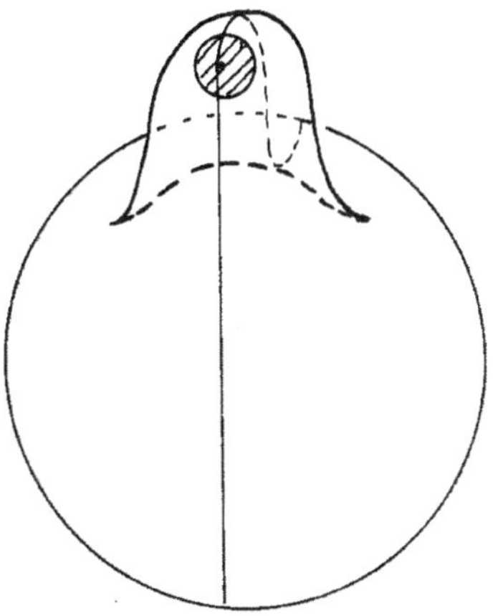

homotopy, we must introduce the singular curve of $\psi_t(u,\cdot)$ with care. Namely, at each point on the singular curve of $\psi_t(u,\cdot)$, $S(\psi_t(u,\cdot))$, the tangent line along the $H_t$-image of the f-level curve through that point cannot coincide with the kernel of $T\psi_t(u,\cdot)$. In an idealized drawing of $K_t(u\times\mathbb{D})$, with some $H_t$ images of f-level curves, we indicate a singular curve that works. The map $\psi_t(u,\cdot)$ is the result of pulling the lower half of the heavily drawn singular curve over the upper curve creating two cusps and a pocket. Once the pocket is formed it can be stretched to cover any point of $\mathbb{R}^2$, so that $K_t(u\times\mathbb{D}_0)$ can

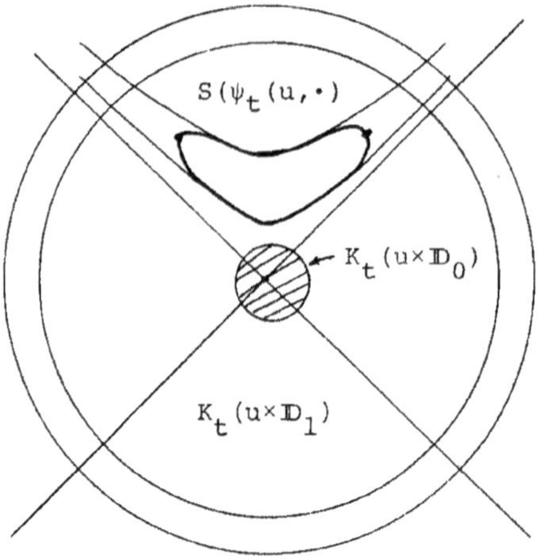


be moved any place on the upper surface of the pocket--any place in $\mathbb{R}^2$.

We apply this technique to each $p_0$, the point in the pair $q^{-1}g(p_0)$, for which $H_t(p_0)$ is not independent of $t$. In this way we have constructed an f-fibre regular homotopy joining $h_0$ to $h_1$ which is fixed on all cusps and double points. // (Lemma)

Our next task is to line up the tangents to the $(f,H_t)$-images of the $S(f)$-curves at all the cusp points. Here is where we use $sp_H \mid \hat{C} = 0$. We work with one cusp, $p_0$, at a time and change $H_t$ in a neighborhood of $p_0$, inside of $\phi(I\times T)$ (where we take $(\phi,\psi,g)$ a $C^2$PN for $f$ at $p_0$).

Since $(f,h_i) \in I_E$, $i = 0, 1$, we know that $H_t(p_0) = 0$ for all $t$. Let $\zeta(t)$ be the unit tangent vector in $\mathbb{R}^2$, which orients $H_t(S(f))$ at $0$. We know that $\zeta(t) = (1,0)$ for $t \in [0,\alpha) \cup (1-\alpha,1]$. Write $\zeta(t) = e^{i\theta(t)}$, where we can take $\theta(0) = 0$. Since $sp_H(p_0) = 0$, we know that $\theta(1) = 0$ as well. Let $\vartheta : \psi(I\times J) \longrightarrow [0,1]$ be a bump function with $\vartheta(f(p_0)) = 1$ and $\vartheta$ vanishing on a neighborhood of $\partial(\psi(I\times J))$. Let $\xi_0$ be any non-zero tangent to $S(f)$ at $p_0$ which orients $S(f)$. We define $\tilde{H}_t$ by:

$$\tilde{H}_t = \begin{cases} H_t & \text{on } M - \phi(I\times T) \\ \{1-\vartheta\circ f+\vartheta\circ f[\,|Th_0(\xi_0)|/|TH_t(\xi_0)|\,]\}e^{-i\theta(t)\vartheta\circ f)}H_t, & \text{on } \phi(I\times T) \end{cases}$$

As before, it is easy to check that $\tilde{H}$ is an f-fibre regular homotopy and that $\tilde{H} : h_0 \simeq h_1$. If $\xi$ is tangent to $S(f)$ at $p_0$, $\xi = \lambda\xi_0$

and

$$\widetilde{TH}_t(\xi) = e^{i\theta(t)}(|Th_0(\xi_0)|/|TH_t(\xi_0)|)\lambda TH_t(\xi_0) = \lambda|Th_0(\xi_0)| = Th_0(\xi)$$

since $TH_t(\xi_0) = |TH_t(\xi_0)|\zeta(t)$ and $|Th_0(\xi_0)| = Th_0(\xi_0) \in \mathbb{R} \subseteq \mathbb{C}$. Thus the $\tilde{H}_t$ image of $S(f)$ at the cusp $p_0$ is tangent to the x-axis and $\widetilde{TH}_t \mid T(S(f))_{p_0} = Th_0 \mid T(S(f))_p$ for all $t$.

Having modified $H$ in this way at every cusp, we now have an f-fibre regular homotopy $H : h_0 \simeq h_1$ which agrees with $E$ on {cusps, double points} and at each point $p \in \hat{C}$, $T(f,H_t) \mid TS(f)_p = TE \mid TS(f)_p$.

For each $t$, the map $(f,H_t) \mid S(f)$ is an immersion of $S(f)$ in $\mathbb{R}^4$ fixed at the double points. Obviously $(1\times f\times H)$ is also an immersion of $I \times S(f)$ in $I \times \mathbb{R}^4$. Consider the homotopy $(1\times f\times(sH+(1-s)h_0) \mid I \times S(f)$. For each $s$ this is again an immersion since at any cusp $p$, $TH \mid TS(f)_p = Th_0 \mid TS(f)_p$. For $t$ in a neighborhood of $\partial I$, this regular homotopy is constant. By an easy variant of the argument that immersions with normal crossings are homotopically stable [$L_1$ or $G^2$] or the triviality of families of embeddings [Lima, P] we obtain an isotopy of the identity of $I \times \mathbb{R}^2 \times \mathbb{R}^2$ with itself, $\Phi_s : I \times \mathbb{R}^2 \times \mathbb{R}^2 \circlearrowleft$ such that $\Phi_s(t,f(x),sH_t(x)+(1-s)h_0(x)) = (t,f(x),h_0(x))$, for $x \in S(f)$, and all $t \in I$. Since our family of immersions of $I \times S(f)$ is constant on a neighborhood of $\partial I \times S(f)$, so also is $\Phi_s$ constant on a neighborhood of $\partial I \times \mathbb{R}^2 \times \mathbb{R}^2$. Thus $\Phi_s(j,u,v) = (u,v)$ for $j = 0, 1$ and for all $s \in [0,1]$. Because the immersions of $I \times S(f)$ is independent of $s$ in its first two components, $\Phi_s(t,u,v) = (t,u,\phi_s(t,u,v))$. Thus we define

$$H^* : I \times M \longrightarrow \mathbb{R}^2 : (t,x) \longrightarrow \phi_1(t,f(x),H_t(x)).$$

As we remarked above $H^*(j,x) = h_j(x)$, $j = 0, 1$. Thus $H^* : h_0 \simeq h_1$, and $H_t^* \mid I \times S(f) = h_0 \mid S(f)$. Since $\Phi_1 \circ (1\times f\times H) \mid t \times M$ is the composition of a diffeomorphism $\Phi_1$ and an immersion $1 \times f \times H$, $(f,H_t^*)$ is an immersion. So $H^*$ is an f-fibre regular homotopy, $H^* : h_0 \underset{E}{\simeq} h_1$ as required. // (injectivity of (E))

## 3.11 The equivalence relation on $\mathcal{Q}$ and the classification theorem

In §3.9 we showed that $[I_\Sigma]_\Sigma$ is in 1:1 correspondence with a collection of functions $\mathcal{Q}$. The set $\mathcal{Q}$ is a product $L \times Z \times G$ where $Z \times G = \underset{R \in R}{\times} (\mathbb{Z}^{2m(R)} \times \mathbb{Z}^{n(R)-1})$ (see §3.8 and §3.9 for the introduction of $Z$ and $G$ respectively) where $m(R)$ is the number of handles in R and $n(R)$ the number of boundary components of R; that is $Z \times G = \underset{R \in R}{\times} H_1(R, \partial R)$. The set $L \subseteq \mathcal{D} \times (C \cup A^*, \mathbb{Z})$, $L = \{\delta \times N^* \mid \delta \in \mathcal{D}\}$ (see §3.4 and §3.8 for $\mathcal{D}$ and $N^*_\delta$ respectively). The first object of this section is to identify the equivalence relation in $\mathcal{Q}$ that corresponds to the projection $P : [I_\Sigma]_\Sigma \longrightarrow [I]$. It will be more convenient to consider $P : [I_V]_V \longrightarrow [I_V]$ as mentioned at the beginning of §3.10.

In §3.7 we introduced a function $\omega_\delta \in (C \cup A^*, \mathbb{R})$ for each $\delta \in \mathcal{D}$, in terms of which we defined $\tau(\delta,\nu) = (\nu - \omega_\delta) \mid C$ and $\sigma(\delta,\nu) = (\nu - \omega_\delta) \mid A^*$ for any $\nu \in (C \cup A^*, \mathbb{Z})$. If $(f,h) \in I_V$ then $\tau(\delta_h, \nu_h) = \tau_h$ and $\sigma(\delta_h, \nu_h) = \sigma_h$.

Given any arc in $C$, there is a unique arc of $S(f) - q^{-1}(V)$ given by $q^{-1}(c) \cap S(f)$. In this section we will consider the elements of $C$ as oriented open arcs in $S(f)$.

An element of $C \cup A^*$ is said to abut a point $v \in V^*$ if $v$ is in its closure. Since all of the arcs of $C \cup A^*$ are oriented if an arc abuts a point $v \in V^*$ it is either arriving, departing or in transit. We will denote the arcs that abut $v$ as $(C \cup A^*)(v) = C(v) \cup A^*(v)$.

Definition. Let $\sim$ be the equivalence relation in $L$ generated by the following: $(\delta,\nu) \sim (\delta',\nu')$ iff $\delta = \delta'$ and $\nu = \nu'$ except on $(C \cup A^*)(v)$ for one $v \in V^*$. If $c \in (C \cup A^*)(v)$, then

$$\nu'(c) - \nu(c) = \begin{cases} +1 & \text{if } c \text{ arrives at } v \\ -1 & \text{if } c \text{ departs at } v \\ 0 & \text{if } c \text{ is in transit through } v \end{cases}$$

Thus we see that if $\nu \sim \nu'$ then for every closed arc $c \in C$, $\nu = \nu'$ on $(C \cup A^*)(v^c)$. We illustrate the generating relations at other points of $V^*$.

1) If $v \in V$, of type $(1{\cdot}2{\cdot}2{\cdot}1)$

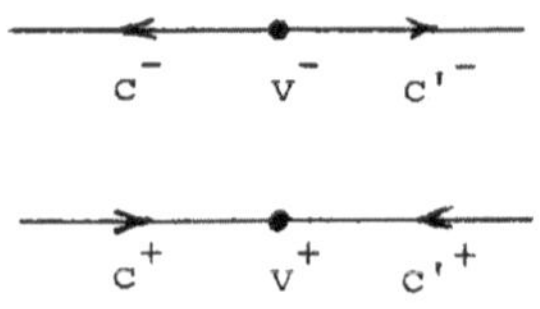


at $v^+$

$$\begin{cases} \nu'(a) = \nu(a) - 1 & \nu'(c'^+) = \nu(c'^+) + 1 \\ \nu'(b) = \nu(b) + 1 & \nu'(c^+) = \nu(c^+) + 1 \\ \nu'(c) = \nu(c) - 1 & \\ \nu'(d) = \nu(d) + 1 & \end{cases}$$

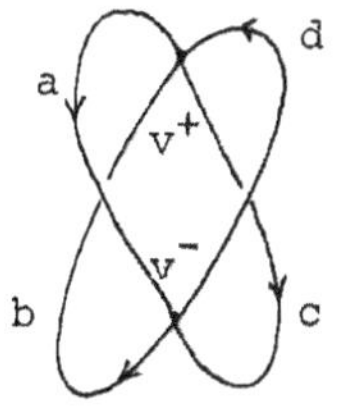


at $v^-$

$$\begin{cases} \nu'(a) = \nu(a) + 1 & \nu'(c^-) = \nu(c^-) - 1 \\ \nu'(b) = \nu(b) - 1 & \nu'(c'^-) = \nu(c'^-) - 1 \\ \nu'(c) = \nu(c) + 1 & \\ \nu'(d) = \nu(d) - 1 & \end{cases}$$

2) If $v \in V$, of type $(1 \cdot 2 \cdot 2 \cdot 3)$

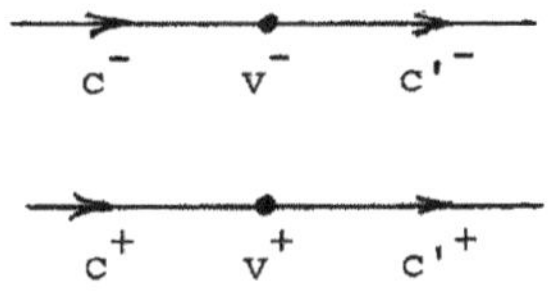


at $v^+$

$$\begin{cases} \nu'(a) = \nu(a) + 1 & \nu'(c^+) = \nu(c^+) + 1 \\ \nu'(b) = \nu(b) & \nu'(c'^+) = \nu(c'^+) - 1 \\ \nu'(c) = \nu(c) - 1 & \end{cases}$$

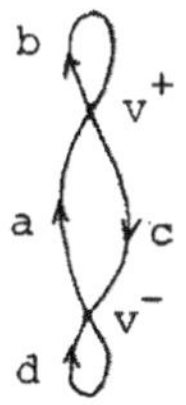


at $v^-$

$$\begin{cases} \nu'(a) = \nu(a) - 1 & \nu'(c^-) = \nu(c^-) + 1 \\ \nu'(c) = \nu(c) + 1 & \nu'(c'^-) = \nu(c'^-) - 1 \\ \nu'(d) = \nu(d) & \end{cases}$$

3) If $v$ is a cusp

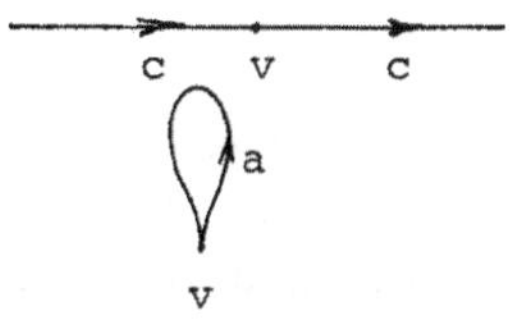


$$\begin{cases} \nu'(a) = \nu(a) \\ \nu'(c) = \nu(c) + 1 \\ \nu'(c') = \nu(c') - 1 \end{cases}$$

We will denote an equivalence class of $(\delta,\nu) \in L$ by $[\delta,\nu]$. Recall that $\lambda : [I_V]_V \longrightarrow L : [f,h]_V \longrightarrow \lambda_h$ was defined by $\lambda_h = (\delta_h,\nu_h)$ (see §3.6).

<u>Proposition 1.</u> <u>$\lambda$ gives a 1:1 correspondence between elements in the fibre of $P$ and the elements in a $\sim$-equivalence class in $L$.</u>

<u>Proof</u>. We first show given $(f,h) \in I_V$, if $(\delta',\nu') \sim \lambda_h$ that we can find an $f$-fibre regular homotopy $H : h \simeq h'$ such that $\lambda_{h'} = (\delta',\nu')$.

To deform $h$ to $h'$ we work entirely in the image $\phi(I\times\mathbb{D})$ where $I \times \mathbb{D} \subseteq I \times T \longrightarrow M$ is part of a $C^2$PN and $\phi(0,0) = v \in V^*$. We introduce a full turn in the $h$-image of the transverse disc $\phi(0\times\mathbb{D})$ about $v$ exactly as we did in the proof of surjectivity in the proof of Theorem 2 of §3.10.

From the form of the deformation used there we see that on an arc of $C$, the value of $\nu$ increases by 1 if the arc arrives at $v$ and decreases by 1 if it leaves $v$. If the arc is in transit at $v$ its $\nu$ value is unchanged. We illustrate:

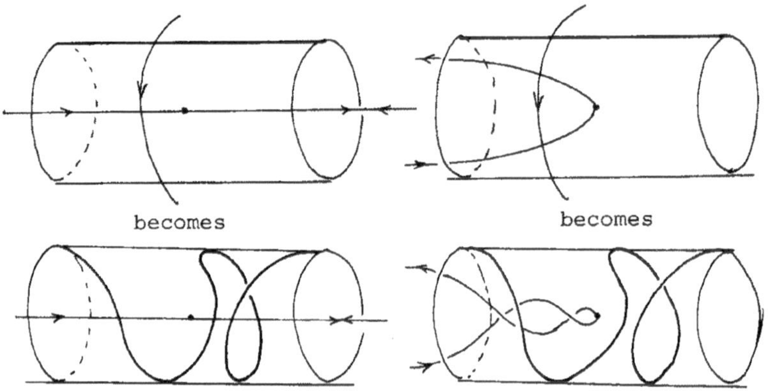


in the non-cusp case in the cusp case

In the figures the arcs with the arrows are the $(f,h)$ images of the arcs of $S(f)$. The heavy dot is the image of $v$. The heavy curve shows the effect of the twist.

To see how this twist affects the values of $\nu$ on the arcs of $A^*(v)$, note that the disc $h(\phi(0\times\mathbb{D})$ is turned a full turn at $h(\phi(0,0))$ and the amount of turning diminishes to zero on $h(\phi(0\times\partial\mathbb{D})$. Thus a ray in $h(\phi(0\times\mathbb{D}))$ through $h(\phi(0,0))$ is deformed to an arc whose tangent image traces a complete circle in the positive sense as we travel along the arc toward the center of the disc. We illustrate the effect of such a twist on a ray in the transverse disc.

Thus we have deformed $h$ to produce an $h'$ so that $\lambda_{h'}$ and $\lambda_h$ satisfy one of the generating relations for $\sim$. Thus we've shown that $\lambda(P^{-1}[f,h])$ contains the equivalence class of $\lambda_h$. To see that $\lambda : P^{-1}[f,h] \longrightarrow L$ is 1:1 we note that there is a homomorphism from $(\hat{V},\mathbb{Z})$ to $(C \cup A^*,\mathbb{Z})$ which is described as follows. Given $s \in (\hat{V},\mathbb{Z})$ its image is $\rho(s)$ where given any arc $c \in C \cup A^*$, $\rho(s)(c)$ is the integer defined as follows: Let $(f,h) \in I_V$ and let $(f,h')$ be the element obtained from $(f,h)$ by spinning the image of the transverse manifold to $v \in \hat{V}$, $s(v)$ full turns. Then $\nu_{h'}(c) - \nu_h(c) = \rho(s)(c)$ defines the mapping $\rho(s)$ uniquely. Notice that if $s : \hat{V} \longrightarrow \mathbb{Z}$ is constant, then $\rho(s) = 0$ since every arc of $C \cup A^*$ joins two not-necessarily-distinct vertices at one of which the arc is arriving and at the other of which the arc is leaving. Thus this map, $\rho$ is well defined and injective

$$\rho : [\hat{V},\mathbb{Z}] \longrightarrow (C \cup A^*,\mathbb{Z}).$$

Obviously, on any arc in $A_{v^c}$ for $c$ closed and on $c \in C$, closed $\rho(s)$ vanishes for all $s \in (V,\mathbb{Z})$. Thus we have a map, for any $(f,h) \in I_V$,

$$\begin{array}{ccccc} P^{-1}[f,h] & \longrightarrow & [\hat{V},\mathbb{Z}] & \longrightarrow & L \\ & & [s] & \longrightarrow & (\delta_h,\nu_h+\rho(s)) \\ [f,h']_V & \longrightarrow & [sp_{(h,h')}] & & \end{array}$$

The composition of these two injective maps is $\lambda$. //

We make the function $\rho(sp_{(h,h')})$ explicit.

<u>Proposition 2</u>. <u>Let</u> $[f,h] = [f,h'] \in [I_V]$ <u>and let</u> $H : h \simeq h'$ <u>be an f-fibre regular homotopy</u>. <u>Let</u> $c \in C \cup A$ <u>which departs from</u> $v_0$ <u>and arrives at</u> $v_1$, $v_i \in \hat{V}$, <u>then</u>

$$\nu_{h'}(c) = \nu_h(c) + sp_H(v_1) - sp_H(v_0).$$

<u>Proof</u>. For any map $k : I \longrightarrow S^1$, $\alpha(k) = \theta(1) - \theta(0)$ where $e^{i\theta(t)} = k(t)$.

<u>Lemma</u>. <u>Let</u> $k_i : I \longrightarrow S^1$, $i = 0, 1$, $I = [0,1]$. <u>Suppose there is a homotopy</u> $K : J \times I \longrightarrow S^1$, $K : k_0 \simeq k_1$, $J = [0,1]$. <u>Then</u>

$$\alpha(k_1(\cdot)) - \alpha(k_0(\cdot)) = \alpha(K_.(1)) - \alpha(K_.(0)).$$

Proof. This is obvious since the degree of $K \mid \partial(J\times I) = 0$, and degree $K \mid \partial(J\times I) = \alpha(k_0(\cdot)) + \alpha(K_.(1)) - \alpha(k_1(\cdot)) - \alpha(K_.(0))$. //

Proof of the Proposition 2. We prove this for $\tau$ and $\sigma$.

1) Recall the definition of $\tau$ (see §3.5). We take a standard vector field along $q^{-1}(c) \cap S(f)$, say $\xi : I \longrightarrow T(M)|_{q^{-1}(c)\cap S(f)}$ such that $Tf\circ\xi$ vanishes. $\tau_h(c)$ is defined as $\alpha(Th\circ\xi/|Th\circ\xi|)$. Let $K = TH\circ\xi/|TH\circ\xi|$, where $K(s,t) = (TH_s(\xi(t))/|TH_s(\xi(t))|$. The lemma gives $0 = \tau_h(c) + \alpha(K_.(1)) - \tau_{h'}(c) - \alpha(K_.(0))$. But $K_.(i) = TH_.(\xi(i))/|TH_.(\xi(i))|$ so $\alpha(K_.(i)) = sp_H(v_i)$ by definition (§3.10).

2) Similarly let $c \in A$. We recall the definition of $\sigma_h(c)$ (§3.6). Let $\rho : [0,1] \longrightarrow TM$ be the arc of unit tangent vectors which orient $c$, then $\sigma_h(c) = \alpha(Th\circ\rho/|Th\circ\rho|)$. Here again letting $K = Th\circ\rho/|TH\circ\rho|$ we get

$$\sigma_{h'}(c) - \sigma_h(c) = sp_{hh'}(v_1) - sp_{hh'}(v_0)$$

all that is needed now is to recall that $\nu_h \mid A^* - \sigma_h$ and $\nu_h \mid C - \tau_h$ depend only on $\delta_h$, and since $[f,h] = [f,h']$, $\delta_h = \delta_{h'}$.

// (Proposition 2 and Proposition 1)

Thus we have shown that $\lambda$ maps the fibre of $P$ over $[f,h]$ bijectively onto the elements of $L$ equivalent to $\lambda_h$.

Consider $\Sigma^* = q^{-1}(V_*) \cup S(f)$ as the 1 dimensional complex whose 1-cells are the components of $q^{-1}(c) \cap S(f) - V^*$ and of $q^{-1}(v) - S(f)$ for $v \in V_*$ and whose vertices are the elements of $V^* = q^{-1}(V_*) \cap S(f)$. For each $(f,h) \in I_V$, $\nu_h$ is a 1-cocycle in $\Sigma^*$ and for all $h'$ such that $[f,h'] = [f,h]$ $\nu_{h'}$ and $\nu_h$ are cohomologous. Thus we have a commutative square:

$$\begin{array}{ccc} [I_V]_V & \xrightarrow{\ \lambda\ } & L \subseteq \mathcal{D} \times C^1(\Sigma^*) \\ P\big\downarrow & & \big\downarrow \\ [I_V] & \xrightarrow{\ [\lambda]\ } & [L] \subseteq \mathcal{D} \times H^1(\Sigma^*) \end{array}$$

In §3.9 we showed that $(\pi,\gamma) : [I_V]_V \longrightarrow L \times Z \times G$ was a 1:1 correspondence. We know that the map $\zeta : [I_V]_V \longrightarrow Z$ is constant on the fibre of $P$ since it is determined by what $(f,h)$ does to the handle generators of $H_1(B(R))$. Thus we have a commutative square

$$\begin{array}{ccc} [I_V]_V & \xrightarrow{(\lambda,\zeta)} & L \times Z \\ \Big\downarrow P & & \Big\downarrow \\ [I_V] & \xrightarrow{[\lambda],\zeta} & [L] \times Z \end{array}$$

where the fibres are in 1:1 correspondence. Here $[L] = L/\sim$. If $\gamma : [I_V]_V \longrightarrow G$ were defined in such a way that it is constant on the fibres of P we would have our final result:

<u>Theorem.</u> $[I_V] \xrightarrow{([\lambda],\zeta,\gamma)} [L] \times Z \times G$ <u>is a 1:1 correspondence.</u>

To get this result we return to the definition of $\gamma$ and reduce the arbitrariness in its construction a little.

In §3.9, to define $\gamma : [I_\Sigma]_\Sigma \longrightarrow G$, we arbitrarily chose representatives of the fibre of $\pi = (\lambda,\zeta) : I_\Sigma \longrightarrow (L\times Z) = P$; $(f,h_p)$ was chosen in $\pi^{-1}(p)$. We defined $\gamma$ by comparing any $(f,h) \in \pi^{-1}(p)$ with $(f,h_p)$. That comparison was possible since for any $(f,h) \in I_\Sigma$, the germ of h at $\hat{\Sigma}$, $(h)_{\hat{\Sigma}}$ was uniquely determined by $\pi(f,h)$ (see §3.6). This uniqueness represented another arbitrary choice: a set of models of germs of mappings at $\hat{\Sigma}$, $H_\Sigma$ was chosen as the image of a section of $\lambda : H_{S,\Sigma} \longrightarrow L_\Sigma$ and $I_\Sigma$ was defined as those elements $(f,h) \in I$ with $(h)_{\hat{\Sigma}} \in H_\Sigma$. We follow precisely the same route in defining $\gamma$ now except that we choose a more convenient set $H_\Sigma \subseteq H_{S,\Sigma}$. If we denote by $h_\ell$ the unique element in our chosen $H_\Sigma$ for which $\lambda(h_\ell) = \ell$, then our requirement on $H_\Sigma$ is:

<u>For any $\ell \sim \ell'$ in $L$ there is a germ of an f-fibre regular homotopy $(H_{\ell\ell'})_{I\times\hat{\Sigma}}$ joining $(h_\ell)_{\hat{\Sigma}}$ and $(h_{\ell'})_{\hat{\Sigma}}$ such that if $h_\ell$ and $h_{\ell'}$ are representatives of the two germs with common domain $B(\Sigma)$, then there is an f-fibre regular homotopy $H^* : h_\ell \simeq h_{\ell'}$ with $(H^*)_{I\times\hat{\Sigma}} = (H_{\ell\ell'})_{I\times\hat{\Sigma}}$ and $H^*$ is constantly $h_\ell$ on $\partial B(\Sigma)$.</u>

That we can choose such a "good" $H_\Sigma$ is easy. The obvious way to make such a choice is to choose one $\ell$ in each equivalence class $[\ell] \in [L]$ and choose $(h_\ell)_{\hat{\Sigma}} \in \lambda^{-1}(\ell)$. For each $\ell' \sim \ell$, construct $(h_{\ell'})_{\hat{\Sigma}}$ from a representative by inserting the appropriate number of twists in the maps in the tubular neighborhoods of $S(f)$ in the transverse manifolds at points of V as was done in §3.10. The twists are spread smoothly along the arcs c of $S(f)$ so that the resulting germ $(h_{\ell'}) \in H_S$. The homotopy inserting the twists gives our required $(H_{\ell\ell'})_{I\times\hat{\Sigma}}$.

For such a "good" choice of $H_\Sigma$, we define $\gamma$ exactly as in §3.9 with the analogous care in choosing representatives $h_p \in I_\Sigma$ in $\pi^{-1}(p)$. Namely for each class $([\ell],z) \in [L] \times Z$ for the same $\ell \in [\ell]$ as we chose above, choose $h_p \in I_\Sigma$ with $p = (\ell,z)$. For all $p' = (\ell',z)$ with $\ell' \sim \ell$, we choose $h_{p'} \in I_\Sigma$ by applying a representative of the spinning homotopy germ $(H_{\ell\ell'})_{\hat{\Sigma}}$ in a neighborhood $B(\Sigma)$ and constant outside of $B(\Sigma)$. Denote by $H_{p,p'}$ this global f-fibre regular homotopy connecting $h_p$ with $h_{p'}$. With these preliminaries restricting our choices we now have:

Lemma. $\gamma : [I_\Sigma]_\Sigma \longrightarrow G$ is constant on classes in $[I_\Sigma]$.

Proof. Let $\pi(f,h) = \pi(f,h_p)$, and let $\pi(f,h') = \pi(f,h_{p'})$ with $[f,h] = [f,h']$. Let $(\ell,z) = \pi(f,h)$ and $(\ell',z') = \pi(f,h')$. We know that $z = z'$ since $\zeta$ is constant on f-regular homotopy classes and $\ell \sim \ell'$ by Proposition 1. By our definition of $H_\Sigma$ we know that there is an f-fibre regular homotopy $H : h \simeq h^*$ such that in a neighborhood of $I \times B(\Sigma)$, $H$ and $H_{p,p'}$ coincide and which are constant outside of a neighborhood of $B(\Sigma)$. Obviously $\pi(f,h^*) = \pi(f,h_{p'}) = p'$.

Since we can form $H \cdot H_{pp'} : I \times (M-B(\Sigma))^2 \longrightarrow \mathbb{R}^2$ a homotopy between $h \cdot h_p$ and $h^* \cdot h_{p'}$ we know that $\gamma(f,h) = \gamma(f,h^*)$. Further we know that $[f,h^*] = [f,h] = [f,h']$ and $\lambda(f,h^*) = \lambda(f,h') = \ell'$. Thus by our Proposition 1, $[f,h^*]_V = [f,h']_V$. But by the Theorem of §3.9 we know that $\gamma$ is constant on classes in $[I_\Sigma]_\Sigma = [I_V]_V$. Thus $\gamma(f,h^*) = \gamma(f,h')$ so $\gamma(f,h) = \gamma(f,h')$. //

As has been remarked, if non-empty, $N_\delta^* \subseteq C^1(\Sigma^*) = (C \cup A^*, \mathbb{Z})$ is the translate of a submodule for each $\delta$. Let $N_\delta \subseteq H^1(\Sigma^*)$ be the image of $N_\delta^*$ dividing out by our equivalence relation $\sim$. Thus identifying $Z \times G$ with $\bigcup_{R \in R} H^1(R,\partial R)$ which we write $H^1(R,\partial R)$ an alternative statement of our result is:

Theorem. The map $([\lambda],\zeta,\gamma) : [I] \longrightarrow D \times H^1(\Sigma^*) \times H^1(R,\partial R)$ gives a 1:1 correspondence between $[I]$ and a subset $(\bigcup_{\delta \times D} (\delta \times N_\delta)) \times H^1(R,\partial R)$ where for each $\delta \in D$ where $N_\delta$ is either empty or a translate of a submodule of $H^1(\Sigma^*)$.

# Appendix



In this appendix we give the proofs of some of the results of the paper. We thought it appropriate to separate them from the main text because of their length or technical nature.

## A-2.1 Lemma 0 and Corollaries 1, 2, 3

*Notation.* If $S_0$, $S_1$ are subsets of a topological space we say $S_0$ *is nested in* $S_1$ or $S_0 \subseteq S_1$ *are nested* if $\overline{S}_0 \subseteq S_1^{\circ}$.

By *an interval* $I$ *in* $\mathbb{R}^m$ we mean the m-fold product of real intervals parallel to the coordinate axes with $0$ in their interiors. An interval $I_1$ in $\mathbb{R}^m$ *is a subinterval* of $I$ if $I_1$ is nested in $I$.

*Remark.* In the following lemma, $T$ is an interval in $\mathbb{R}^p$, and the metric space in which everything is defined is $T \times \mathbb{R}^q \times \mathbb{R}^r \times \mathbb{R}^n$; all topological notions refer to this space and *not* to the containing $\mathbb{R}^m$, $m = p + q + r + n$.

*Lemma 0. Let* $T$ *be an interval of* $\mathbb{R}^p$, *and* $J$ *and* $K$ *be compact intervals of* $\mathbb{R}^q$ *and* $\mathbb{R}^r$ *respectively. Let* $N$ *be a closed neighborhood of* $T \times J \times K \times \{0\}$ *in* $T \times J \times K \times \mathbb{R}^n$, *and let* $f : N \longrightarrow \mathbb{R}^n$ *define an orientation preserving diffeomorphism,* $1 \times f : N \longrightarrow T \times J \times K \times \mathbb{R}^n$ *such that*

a) $f \mid T \times J \times K \times \{0\} = 0$.

b) *For* $J_0$ *a closed subinterval of* $J$, $f \mid N \cap (T \times J - J_0 \times K \times \mathbb{R}^n) =$ proj, *projection onto the* $\mathbb{R}^n$ *factor.*

c) *For some* $(t_0, v_0) \in T \times K$, $\mathrm{Jac}_x f(t_0, \cdot, v_0, 0) : J \longrightarrow GL(n, \mathbb{R})$ *is homotopic to the map constantly the identity,* $I \in GL(n, \mathbb{R})$ *such that the homotopy is constantly* $I$ *on* $J - J_0$. *(Here* $\mathrm{Jac}_x f(t_0, u, v_0, 0)$ *is the Jacobian matrix of* $f \mid (\{t_0, u, v_0\} \times \mathbb{R}^n) \cap N$ *evaluated at* $0 \in \mathbb{R}^n$*), then for any closed subintervals* $J_0 \subseteq J_1 \subseteq J$ *and* $K_0 \subseteq K_1 \subseteq K$ *and sufficiently small closed nested neighborhoods of* $T \times J \times K \times \{0\}$, $N_0 \subseteq N_1 \subseteq N$ *there are orientation preserving diffeomorphisms:* $1 \times g : T \times \mathbb{R}^q \times \mathbb{R}^r \times \mathbb{R}^n \longrightarrow T \times \mathbb{R}^q \times \mathbb{R}^r \times \mathbb{R}^n \longleftarrow N : 1 \times h$ *such* *that* $g \mid T \times \mathbb{R}^q \times \mathbb{R}^r \times \{0\} = 0 = h \mid T \times J \times K \times \{0\}$, *and*

1) *On* $N_0 \cap (T\times J\times K_0\times \mathbb{R}^n)$, $g = f$, $h = \text{proj}$.

2) *On* $T \times \mathbb{R}^q \times \mathbb{R}^r \times \mathbb{R}^n - N_1 \cap (T\times J_1\times K_1\times \mathbb{R}^n)$, $g = \text{proj}$ *and on* $N - N_1' \cap (T\times J_1\times K_1\times \mathbb{R}^n)$, $h = f$, *where* $N_1' = (1\times f)^{-1}(N_1)$.

*Remark*. If $T$ is also assumed compact, the neighborhoods, $N_0$ and $N_1$ may be taken in the form $T \times J \times K \times B_i$ where $B_0 \subseteq B_1$ are nested balls in $\mathbb{R}^n$.

We state two special cases that are used a number of times in the sequel.

If $q = r = 0$:

*Corollary 1. Let* $T$ *be an interval in* $\mathbb{R}^p$. *Let* $N$ *be a neighborhood of* $T \times \{0\} \subseteq T \times \mathbb{R}^n$. *Let* $f : N \longrightarrow \mathbb{R}^n$ *define an orientation preserving diffeomorphism* $1 \times f : N \longrightarrow T \times \mathbb{R}^n$ *such that* $f \mid T \times \{0\} = 0$, *then for sufficiently small closed nested neighborhoods of* $T \times \{0\}$, $N_0 \subseteq N_1 \subseteq N$ *there is an orientation preserving diffeomorphism* $1 \times g$ *of* $T \times \mathbb{R}^n$ *such that* $g \mid T \times \{0\} = 0$, *and*

(a) *On* $N_0$, $g = f$.

(b) *On* $(T\times \mathbb{R}^n) - N_1$, $g = \text{proj}$.

If $p = q = 0$:

*Corollary 2. Let* $K$ *be a compact interval of* $\mathbb{R}^r$, *and* $B$ *a closed ball about* $0$ *in* $\mathbb{R}^n$. *Let* $1 \times f : K \times B \longrightarrow K \times \mathbb{R}^n$ *be an orientation preserving diffeomorphism with* $f \mid K \times \{0\} = 0$. *Then for sufficiently small closed nested balls,* $B_0 \subseteq B_1 \subseteq B \subseteq \mathbb{R}^n$ *and any closed nested intervals* $K_0 \subseteq K_1 \subseteq K \subseteq \mathbb{R}^r$ *there is an orientation preserving diffeomorphism* $1 \times g : \mathbb{R}^r \times \mathbb{R}^n \longrightarrow \mathbb{R}^r \times \mathbb{R}^n$ *such that* $g \mid \mathbb{R}^r \times \{0\} = 0$ *and:*

(a) *On* $K_0 \times B_0$, $g = f$.

(b) *On* $\mathbb{R}^n \times \mathbb{R}^r - (K_1\ B_1)$, $g = \text{proj}$.

The following is a corollary of the proof and not a formal consequence of the statement of the lemma itself.

*Corollary 3. Let* $T_0 \subseteq T_1 \subseteq T$ *be a nest of intervals in* $T \subseteq \mathbb{R}^p$ *and let* $N$ *be a neighborhood of* $T \times \{0\}$ *in* $T \times \mathbb{R}^n$. *Let* $f : N \longrightarrow \mathbb{R}^n$ *define an orientation preserving diffeomorphism* $1 \times f : N \longrightarrow T \times \mathbb{R}^n$ *such that* $f \mid T \times \{0\} = 0$. *Then for sufficiently small neighborhoods* $N_0$ of $T_0 \times \{0\}$ *and* $N_1$ *of* $T_1 \times \{0\}$, there is an orientation *preserving diffeomorphism* $1 \times g$ *of* $T \times \mathbb{R}^n$ *such that*

$g \mid T \times \{0\} = 0$ <u>and</u>

(a) <u>On</u> $N_0$, $g = f$.

(b) <u>On</u> $T \times \mathbb{R}^n - N_1$, $g = \text{proj}$.

To get this corollary the proof must be modified by making bump functions depend on $t \in T$ as well as on $x \in \mathbb{R}^n$.

<u>Proof of Lemma 0.</u> It suffices to prove the existence of such a $g$ since $1 \times h = (1\times g)^{-1} \circ (1\times f)$ gives the desired $h$.

Write $L = T \times J \times K$ and let $A : L \longrightarrow GL(n,\mathbb{R})$ be given by $A(\omega) = \text{Jac}_x f(\omega,0)$. Define $\tilde{f}$ by $1 \times f = (1\times A) \circ (1\times \tilde{f})$ where $(1\times A)(\omega,z) = (\omega, A(\omega).z)$. Both maps $1 \times A$ and $1 \times \tilde{f}$ satisfy the hypotheses of the lemma. Let $g_A$ ang $g_{\tilde{f}}$ be the maps given by the lemma; it is no restriction to assume that the same sets, $J_1$, $K_0$, $K_1$, $N_0$, $N_0$ are used for both maps. If we let $N_0' = N_0 \cap (1\times\tilde{f})^{-1}(N_0)$ and leave all the other sets unaltered we get our result for $f$ setting $1 \times g_f = (1\times g_A) \circ (1\times g_{\tilde{f}})$. On $N_0' \cap (T\times J\times K_0 \times \mathbb{R}^n)$, $1 \times g_{\tilde{f}} = 1 \times \tilde{f}$ since $N_0' \subseteq N_0$. Since $(1\times\tilde{f})(N_0') \subseteq N_0$, we see that on $N_0' \cap (T\times J\times K_0\times\mathbb{R}^n)$, $(1\times g_A)\circ(1\times g_{\tilde{f}}) = (1\times A)\circ(1\times\tilde{f}) = 1 \times f$. This gives the first conclusion of Lemma 0 for the composition of $(1\times A)$ and $(1\times\tilde{f})$. The second conclusion is simpler. On $(T\times J\times K\times\mathbb{R}^n) - (N_1 \cap (T\times J_1\times K_1\times\mathbb{R}^n))$ both maps $1 \times g_A$ and $1 \times g_{\tilde{f}}$ are the identity hence so is their composition. Thus it suffices to prove the lemma in two cases: <u>Case 1</u> in which $\text{Jac}_x f(\omega,0) = I$ and <u>case 2</u> where $f(\omega,z) = A(\omega)z$, where $A : L \longrightarrow GL(n,\mathbb{R})$.

<u>Case 1.</u> $\text{Jac}_x f(\omega,0) = I$ for all $\omega \in L$.

Let $\vartheta : \mathbb{R} \longrightarrow [0,1]$ be a smooth function such that $\vartheta \mid (-\infty,0] = 1$ and $\vartheta[1,\infty) = 0$, and $0 \geqslant \vartheta' \geqslant -2$. For any $0 < \sigma < \tau$, let

$$\psi_{\sigma,\tau} : \mathbb{R}^n \longrightarrow [0,1] : x \longrightarrow \vartheta\left(\frac{|x|^2 - \sigma^2}{\tau^2 - \sigma^2}\right),$$

so

$$\psi_{\sigma,\tau}(x) = \begin{cases} 1 & \text{if } x \in B_\sigma \\ 0 & \text{if } x \notin B_\tau \end{cases}$$

($B_\rho$ is the closed ball or radius $\rho$ about 0.)

Assuming $\sigma^2 \leqslant \tau^2/2$ we have:

$$\left|\frac{\partial\psi_{\sigma,\tau}}{\partial x_i}(x)\right| = \left|\frac{2x_i}{\tau^2-\sigma^2}\right|\left|\vartheta'\left(\frac{|x|^2-\sigma^2}{\tau^2-\sigma^2}\right)\right| \leqslant \frac{4\tau}{\tau^2-\sigma^2} \leqslant \frac{8}{\tau}$$

Similarly, if $K_0 \subseteq K_1 \subseteq K$ are nested intervals in $\mathbb{R}^r$ let $\phi : \mathbb{R}^r \longrightarrow [0,1]$ such that $\phi \mid K_0 = 1$ and $\phi \mid \mathbb{R}^n - K_1 = 0$.

Now write $f(\omega,x) = x + h(\omega,x)$, where $h(\omega,0) = 0$, and $\mathrm{Jac}_x h(\omega) = 0$. Define two functions $\sigma, \tau : T \times J \times K \longrightarrow (0,\infty)$ such that $0 < \sigma < \tau$ and $N_0 = \{(\omega,x) \mid |x| \leqslant \sigma(\omega)\} \subseteq N_1 = \{(\omega,x) \mid |x| \leqslant \tau(\omega)\} \subseteq N$. A more stringent requirement will be placed on $\tau$ at the end of the proof.

Define $k : T \times \mathbb{R}^q \times \mathbb{R}^r \times \mathbb{R}^n \circlearrowleft$ by

$$k = \begin{cases} h\cdot\phi\cdot\psi_{\sigma,\tau} & \text{on } N \\ 0 & \text{on } (T\times\mathbb{R}^q\times\mathbb{R}^r\times\mathbb{R}^n) - N \end{cases}$$

This is obviously well defined since $\psi_{\sigma,\tau}$ vanishes on $N - N_1$. Let $g(\omega,x) = x + k(\omega,x)$. On $N_0 \cap (T\times J\times K_0\times\mathbb{R}^n)$, $k = h$, which gives conclusion 1. For conclusion 2, note that by its definition $k$ vanishes on the complement of $N_1$ and on the complement of $T \times \mathbb{R}^q \times K_1 \times \mathbb{R}^n$. By hypothesis b), $h$ vanishes on $N \cap (T\times J - J_0\times K\times\mathbb{R}^n)$. Thus $k$ vanishes on the complement of $N_1 \cap (T\times J_0\times K_1\times\mathbb{R}^n)$. Now we must check that $1 \times g$ is a diffeomorphism. We do that by showing that $|\mathrm{Jac}_x k(\omega,x)| < 1/2$ for all $x \in \mathbb{R}^n$. Since $k(\omega,0) = 0 = \mathrm{Jac}_x k(\omega,0)$, this estimate is enough to guarantee that $x + k(\omega,x)$ defines a diffeomorphism of $\mathbb{R}^n$ for each $\omega \in T \times \mathbb{R}^q \times \mathbb{R}^r$ [Lang, p. 12].

$$\left|\frac{\partial k}{\partial x_i}\right| \leqslant \left|\phi\frac{\partial\psi_{\sigma,\tau}}{\partial x_i}h\right| + \left|\phi\psi_{\sigma,\tau}\frac{\partial h}{\partial x_i}\right|$$

This gives:

$$\left|\frac{\partial k}{\partial x_i}\right| \leqslant \begin{cases} 0 & \text{on } (T\times\mathbb{R}^q\times\mathbb{R}^r\times\mathbb{R}^n) - N_1 \cap (T\times J_0\times K_1\times\mathbb{R}^n) \\ \left|\frac{\partial h}{\partial x_i}\right| & \text{on } N_0 \\ \frac{8|h|}{\tau} + \left|\frac{\partial h}{\partial x_i}\right| & \text{on } N_1 - N_0 \end{cases}$$

Since $h(\omega,0) = 0$ and $\mathrm{Jac}_x h(\omega,0) = 0$, we can choose $\tau$ so small that for $0 < |x| < \tau(\omega)$,

$$\frac{8|h(\omega,x)|}{\tau(\omega)} < \frac{8|h(\omega,x)|}{|x|} < \frac{1}{4\sqrt{n}} \quad \text{and} \quad \left|\frac{\partial h}{\partial x_i}(\omega,x)\right| < \frac{1}{4\sqrt{n}}$$

Thus with this choice of $\tau$, on all of $T \times \mathbb{R}^q \times \mathbb{R}^r \times \mathbb{R}^n$,

$$\left|\frac{\partial k}{\partial x_i}(\omega,x)\right| < \frac{1}{2\sqrt{n}} \quad \text{or} \quad |\mathrm{Jac}_x k(\omega,x)| < \tfrac{1}{2}.$$

This is enough to guarantee that $x + k(\omega,x)$ is a diffeomorphism of $\mathbb{R}^n$ for each $\omega \in T \times \mathbb{R}^q \times \mathbb{R}^r$. //

Case 2. $f : N \longrightarrow \mathbb{R}^n : (t,u,v,x) \longrightarrow A(t,u,v)x$. Since $N$ is a neighborhood of $T \times J \times K \times \{0\}$, such an $f$ is obviously defined on all of $L \times \mathbb{R}^n$ by $A : L \longrightarrow GL(n,\mathbb{R})$. By hypothesis b), $A \mid T \times J - J_0 \times K = I \in GL(n,\mathbb{R})$, so we can extend $A$ as $I$ to all of $T \times \mathbb{R}^q \times K$. We will now define $\tilde{A}$ on $T \times \mathbb{R}^q \times \mathbb{R}^r$ so that $A = \tilde{A}$ on $T \times \mathbb{R}^q \times K_0$ and $\tilde{A} = I$ on $T \times \mathbb{R}^q \times (\mathbb{R}^r - K_1)$ for closed nested $K_0 \subseteq K_1 \subseteq K$, and $\tilde{A} = I$ on $T \times (\mathbb{R}^q \times \mathbb{R}^r - J_0 \times K_1)$. Choose closed nested intervals, $K_0 \subseteq K_{1/2} \subseteq K_1 \subseteq K$ and define $\vartheta_i : \mathbb{R}^r \longrightarrow [0,1]$, $i = 1, 2$ so that

$$\vartheta_1 = \begin{cases} 0 & \text{on } K_0 \\ 1 & \text{on } \mathbb{R}^r - K_{1/2} \end{cases} \qquad \vartheta_2 = \begin{cases} 0 & \text{on } K_{1/2} \\ 1 & \text{on } \mathbb{R}^r - K_1 \end{cases}$$

Let $H_s : \mathbb{R}^q \longrightarrow GL(n,\mathbb{R})$ be the homotopy given by hypothesis c), namely. $H_0(u) = A(t_0,u,v_0)$ and $H_1(u) = I$, $u \in \mathbb{R}^q$, and $H_s \mid \mathbb{R}^q - J_0 = I$ for all $s \in [0,1]$. We define $\tilde{A} : T \times \mathbb{R}^q \times \mathbb{R}^r \longrightarrow GL(n,\mathbb{R})$ by:

$$\tilde{A}(t,u,v) = \begin{cases} A(t,u,v) \quad , \quad \text{if } v \in K_0 \\ A(\vartheta_1(v)t_0+(1-\vartheta_1(v))t,u,\vartheta_1(v)v_0-(1-\vartheta_1(v))v), \text{ if } v \in K_{1/2}-\mathring{K}_0 \\ H_{\vartheta_2(v)}(u) \quad , \quad \text{if } v \in \mathbb{R}^r - \mathring{K}_{1/2} \end{cases}$$

The map $\tilde{f} : T \times \mathbb{R}^q \times \mathbb{R}^r \times \mathbb{R}^n \longrightarrow \mathbb{R}^n : (t,u,v,x) \longrightarrow \tilde{A}(t,u,v)x$ satisfies $\tilde{f}(t,u,v,x) = A(t,u,v)x$ for $(t,u,v) \in T \times \mathbb{R}^q \times K_0$ and

a) $\tilde{f} \mid T \times \mathbb{R}^q \times \mathbb{R}^r \times \{0\} = 0$.

b) For the closed subinterval $J_0 \times K_1 \subseteq \mathbb{R}^q \times \mathbb{R}^r$, $\tilde{f} \mid T \times (\mathbb{R}^q \times \mathbb{R}^r - J_0 \times K_1) \times \mathbb{R}^n = \mathrm{proj}$.

c) $\tilde{A}(t_0,u,v)$ is homotopic to $I$ in such a way that it is constantly $I$ on $(\mathbb{R}^q \times \mathbb{R}^r - J_0 \times K_1)$.

The only one of these properties of $\tilde{A}$ that needs checking is c). To that end let:

$$K_s(u,v) = \begin{cases} \begin{cases} 2s\tilde{A}(t_0,u,v_0) + (1-2s)\tilde{A}(t_0,u,v), & 0 \leq s \leq 1/2 \\ H_{2s-1}(u), & 1/2 \leq s \leq 1 \end{cases}, & v \in K_{1/2} \\ \begin{cases} H_{\vartheta_2(v)}(u), & 0 \leq s \leq 1/2 \\ H_{2s-1+\vartheta_2(v)(2-2s)}(u), & 1/2 \leq s \leq 1 \end{cases}, & v \in \mathbb{R}^r - K_{1/2} \end{cases}$$

This gives the required homotopy.

Thus the map $\tilde{f}$ given by $\tilde{A}$ satisfies the original hypotheses with $J \times K$ and $J_0 \times K_1$ playing the roles of $J$ and $J_0$ respectively. Thus it is enough to prove the following slightly strengthened special case of Lemma 0.

Lemma 0'. Given an interval $T$ in $\mathbb{R}^p$ and a compact interval $J$ in $\mathbb{R}^q$ if $A : T \times \mathbb{R}^q \longrightarrow SL(n,\mathbb{R})$ satisfies:

b') For a closed subinterval $J_0 \subseteq J$, $A \mid T \times (\mathbb{R}^q - J_0) = I$

c') For some $t_0 \in T$, there is a homotopy $H_s : A(t_0,\cdot) \simeq I$ such that $H_s \mid \mathbb{R}^q - J_0 = I$ for all $s \in [0,1]$.

Then for a sufficiently small closed ball $B_1 \subseteq \mathbb{R}^n$, there is a closed neighborhood $N_0$ of $T \times \mathbb{R}^q \times \{0\}$ such that $N_0 \subseteq T \times \mathbb{R}^q \times \mathring{B}_1$ and for any closed nested intervals $J_0 \subseteq J_1 \subseteq J$, there is a diffeomorphism $1 \times g : T \times \mathbb{R}^q \times \mathbb{R}^n \longrightarrow T \times \mathbb{R}^q \times \mathbb{R}^n$ such that

1') On $N_0$, $g(t,u,x) = A(t,u)x$

2') On $T \times (\mathbb{R}^q \times \mathbb{R}^n - J_1 \times B_1)$, $g(t,u,x) = x$.

Note. Conclusion 1') is stronger than the original conclusion 1).

Proof of Lemma 0'. We write $A$ in its polar decomposition, $A = P\cdot\mathcal{O}$, where $P$ is positive definite and $\mathcal{O}$ is special orthogonal. Of course wherever $A = I$, then $P = \mathcal{O} = I$ also. Choose a closed interval $J_1$ so that $J_0 \subseteq J_1 \subseteq J$ are nested and let $m : T \times \mathbb{R}^q \longrightarrow \mathbb{R}$ be a smooth function whose value is $1$ on $T \times (\mathbb{R}^q - J_1)$ and such that generally $m(t,u) \geq \max\{1, \text{largest eigenvalue of } P(t,u)\}$. Write $P = m\cdot Q$. We break the proof of Lemma 0' into three cases $A = \mathcal{O}$, $Q$ and $m\cdot I$.

a) $A = \mathcal{O} : T \times \mathbb{R}^q \longrightarrow SO(n)$, where $A = I$ on $T \times (\mathbb{R}^q - J_0)$. Let $\Phi : [0,1] \times T \times \mathbb{R}^q \longrightarrow SO(n)$ be a smooth map such that $\Phi_0 = \mathcal{O}$, $\Phi_1 = \mathrm{id}$ and $\Phi_s \mid T \times (\mathbb{R}^q - J_0) = I$ for all $s \in [0,1]$. To construct $\Phi$ we use the hypothesis c) of the lemma: for the first part of the interval just compose $A$ with the contraction of $T \times \mathbb{R}^q$ to $t_0 \times \mathbb{R}^q$, then use the homotopy $H$ to connect $A(t_0,\cdot)$ with $I$.

Now choose $B_0 \subseteq B_1$ closed nested balls and let $\phi : \mathbb{R}^n \longrightarrow [0,1]$ such that

$$\phi = \begin{cases} 0 & \text{on } B_0 \\ 1 & \text{on } \mathbb{R}^n - B_1 \end{cases}, \quad \text{where } \phi(x) \text{ depends only on } |x|,$$

and define $g : T \times \mathbb{R}^q \times \mathbb{R}^n \longrightarrow \mathbb{R}^n : (t,u,x) \longrightarrow \mathcal{O}_{\phi(x)}(t,u)x$. For $x \in B_0$, $g(t,u,x) = \mathcal{O}(t,u)x$ and if $x \notin B_1$, $g(t,u,x) = x$. Further if $u \in \mathbb{R}^q - J_0$, $g(t,u,x) = x$. Thus we need only check that $1 \times g$ is a diffeomorphism. Letting $h(t,u,x) = \mathcal{O}^T_{\phi(x)}(t,u)x$, we show that $1 \times h$ is $(1\times g)^{-1}$.

$$h(t,u,g(t,u,x)) = \mathcal{O}^T_{\phi(g(t,u,x))}(t,u)\mathcal{O}_{\phi(x)}(t,u)x.$$

But $|g(t,u,x)| = |x|$ so $\phi(g(t,u,x)) = \phi(x)$. $//_{(a)}$

(b) $A = Q : T \times \mathbb{R}^q \longrightarrow \{$positive definite $n \times n$ matrices$\}$, and $Q \mid T \times \mathbb{R}^q - J_1 = I$, $Q \mid T \times (\mathbb{R}^q - J_0) = \frac{1}{m}I$, and generally all eigenvalues of $Q$ are no greater than one. We use the following two properties of the set of real symmetric matrices.

<u>Let $S$ be the manifold of real $(n\times n)$-symmetric matrices,</u> $n \geq 2$ <u>and let $S'$ be the closed subset of $S$ of those matrices with repeated eigenvalues.</u>

1) $\operatorname{codim}_S S' = 2$

2) <u>If $R = S - S'$, there is a smooth map</u> $O : R \longrightarrow O(n)$ <u>such that</u> $O(A)AO(A)^T = \delta(d_1,\ldots,d_n)$, <u>the diagonal matrix with entries</u> $d_i < d_{i+1}$, $i = 1,\ldots,n-1$.

The proofs of 1) and 2) are included since I don't know a convenient reference:

1): Let $\Delta$ be the space of diagonal matrices which we identify with $\mathbb{R}^n$. $O(n)$ acts on $\Delta$ by conjugation, $\alpha : O(n) \times \Delta \longrightarrow S : (O,\delta) \longrightarrow O\cdot\delta\cdot O^T$. We know this is a surjective map. Let $\Delta' = \{\delta(d_1,\ldots,d_n) \in \Delta \mid d_1 = d_2\}$. Then $\alpha(O(n)\times\Delta') = S'$. The isotropy subgroup of $\Delta'$ is just $O(2) \times \mathrm{id}_{\mathbb{R}^{n-2}}$. Thus we have an induced map $\alpha' : \left(O(n)/O(2)\times\mathrm{id}_{\mathbb{R}^{n-2}}\right) \times \Delta' \longrightarrow S$ with image $S'$.

The original map $\alpha$ has rank equal to $\dim(O(n)\times\Delta) = \dim(S)$ on an open-dense set, namely $O(n) \times \Delta_0$, $\Delta_0 = \{\delta(d_1,\ldots,d_n)$ with $d_i \neq d_j$, $i \neq j\}$. Thus the image of $\alpha \mid O(n) \times \Delta'$ has codim at least 2 in $S$, since $\dim\left((O(n)/O(2)\times\mathrm{id}_{\mathbb{R}^{n-2}})\times\Delta_0'\right) = \dim(O(n)\times\Delta) - 2$. On

$((O(n)/O(2)\times \mathrm{id}_{\mathbb{R}^{n-2}})\times\Delta_0')$, where $\Delta_0' = \{\delta(d_1,d_1,d_3,\ldots,d_n) \mid d_i \neq d_j,$ $i \neq j\}$, rank $(\alpha') = \dim(O(n)\times\Delta) - 2$. Thus $\alpha(O(n)\times\Delta')$ has codim 2 in $S$. $//_{1)}$

2): Let $D = \{d \in \mathbb{R}^n \mid d_1 < d_2 < \cdots < d_n\}$. Let $d : R \longrightarrow D :$ $A \longrightarrow d(A)$, the ordered sequence of eigenvalues of $A$. To see that this is a smooth map. The map $p : R \longrightarrow P_n : A \longrightarrow p(A) =$ $\det(A - t\,\mathrm{id}_{\mathbb{R}^n})$ is obviously smooth. We identify $P_n$, the monic polynomials of degree $n$ in $t$ with $\mathbb{R}^n$ by their sequence of coefficients. Define $\sigma : D \longrightarrow \mathbb{R}^n : d \longrightarrow (\sigma_1(d),\ldots,\sigma_n(d))$, where $\sigma_j$ is the $j^{th}$ symmetric function of $n$ variables.

Lemma. $\sigma : D \longrightarrow \mathbb{R}^n$ is a diffeomorphism onto its image.

Proof. The map is obviously 1:1 since the set of roots of an $n^{th}$ degree real polynomial with $n$ real roots is a well defined set. Denote by $\sigma_{i,\hat{j}}$ the restriction of $\sigma_i$ to the set $(d_j = 0)$ in $D$. The Jacobian of $\sigma$ is just:

$$J(d_1,\ldots,d_n) = \begin{pmatrix} 1 & \cdots & 1 \\ \sigma_{1,\hat{1}} & \cdots & \sigma_{1,\hat{n}} \\ \vdots & & \vdots \\ \sigma_{n-1,\hat{1}} & & \sigma_{n-1,\hat{n}} \end{pmatrix}(d_1,\ldots,d_n).$$

To show that $J$ is non-singular we proceed by induction on $n$: For $n = 1$, $J(d_1) = 1$ and for $n = 2$, $J(d_1,d_2) = \begin{pmatrix} 1 & 1 \\ d_2 & d_1 \end{pmatrix}$ which is non-singular since $d_1 \neq d_2$. In general subtract the first column of $J$ from each of the others. The resulting $(i+1,k)^{th}$ entry for $k \geq 2$, $i \geq 1$ is:

$$(\sigma_{i,\hat{k}} - \sigma_{i,\hat{1}})(d) = \sum_{\substack{|\alpha|=i \\ k\notin\alpha}} d^{\alpha} - \sum_{\substack{|\alpha'|=i \\ 1\notin\alpha'}} d^{\alpha'} = (d_1 - d_k) \sum_{\substack{|\beta|=i-1 \\ 1,k\notin\beta}} d^{\beta}$$

Factoring $(d_1 - d_k) \neq 0$ from the $k^{th}$ column for $k \geq 2$ yields

$$\begin{pmatrix} 1 & 0 \quad\quad 0 \\ * & J(d_2,\ldots,d_n) \end{pmatrix} \qquad //_{\text{Lemma}}$$

Since the image of $p : R \longrightarrow P_n = \mathbb{R}^n$ is the same as that of $\sigma : D \longrightarrow \mathbb{R}^n$, we can define the smooth map $d : \sigma^{-1} \circ p : R \longrightarrow D$. Let $\delta(d_1,\ldots,d_n)$ be the diagonal matrix with $d_1,\ldots,d_n$ on the diagonal. Solving for $O \in O(n)$ so that:

$$AO = O\delta(d(A))$$

gives a smooth well defined map from $R$ to $O(n)$ if we require that the first non-zero entry in each column be positive. This is clear since for each $i$, $(A-d_i(A)I)$ has rank $(n-1)$, so there is a unique non-zero unit vector $O_i \in \mathbb{R}^n$ such that $(A-d_i(A)I)O_i = 0$ whose first non-zero entry is positive. The entries in $O_i$, being algebraic functions of those of $(A-d_i(A)I)$, are smooth functions of $A$. Thus the matrix $O$ whose $i^{th}$ column is $O_i$ is smoothly dependent on $A$. // 2)

Continuing the proof of b): Choose any set of nested closed balls $B_0 \subseteq B_1 \subseteq B_0' \subseteq B_1' \subseteq \mathbb{R}^n$. We will work with $B_0'$, $B_1'$ in the second part of the proof only. Let $\phi : \mathbb{R}^n \longrightarrow [0,1]$ be such that $\phi \mid B_0 = 0$ and $\phi \mid \mathbb{R}^n - B_1 = 1$. By virtue of property 1) we can approximate $Q$ by $Q' : T \times \mathbb{R}^q \longrightarrow R$ so well that the map

$$T \times \mathbb{R}^q \times \mathbb{R}^n \ni (t,u,x) \longrightarrow$$

$$(t,u,[Q(t,u)+\phi(x)(Q'(t,u)-Q(t,u))]x) \in T \times \mathbb{R}^q \times \mathbb{R}^n$$

is a diffeomorphism. Property 2) gives us two mappings: $O : T \times \mathbb{R}^q \longrightarrow O(n)$ and $\delta' : T \longrightarrow \mathbb{R}^q \longrightarrow D$, such that $Q' = O\delta' O^T$, where $D$ is the set of diagonal matrices with entries $d_i < d_{i+1}$, $i = 1,\ldots,n-1$. We choose the approximation $Q'$ so that its eigenvalues are no greater than one anywhere; this is possible since the eigenvalues of $Q$ are bounded by one everywhere.

Choose a closed interval $J'$, so that $J_0 \subseteq J' \subseteq J_1$ are nested and define $\delta^\dagger : T \times \mathbb{R}^q \longrightarrow D$ so that

$$\begin{cases} \delta^\dagger = \delta' & \text{on } T \times J_0 \\ \delta^\dagger = \frac{1}{m}I & \text{on } T \times (\mathbb{R}^q - J') \end{cases}$$

and so that on $T \times (\mathbb{R}^q - J_0)$, we make the diagonal entry functions $d_i^\dagger$ and $d_i'$ of $\delta^\dagger$ and $\delta'$ respectively satisfy: $|\frac{1}{m} - d_i^\dagger| \leq |\frac{1}{m} - d_i'|$. Using the function $O$ obtained above we define $Q^\dagger = O\delta^\dagger O^T$. Since $Q \mid T \times (\mathbb{R}^q - J_0) = \frac{1}{m}I$, we see that on $T \times (\mathbb{R}^q - J_0)$:

$$|Q-Q^{\dagger}| = |\tfrac{1}{m}I - O\delta^{\dagger}O^{T}| = |\tfrac{1}{m}I - \delta^{\dagger}| \leqslant |\tfrac{1}{m}I - \delta'| = |Q-Q'|.$$

Since $1 \times (Q+\phi(Q'-Q))$ is a diffeomorphism on $T \times \mathbb{R}^q \times \mathbb{R}^n$, so also is $1 \times (Q+\phi(Q^{\dagger}-Q))$. Let $g^{\dagger} = Q + \phi(Q^{\dagger}-Q)$, then

1) On $(T\times\mathbb{R}^q\times B_0)$, $g^{\dagger}(t,u,x) = Q(t,u)x$.

2) On $(T\times(\mathbb{R}^q-J_1)\times\mathbb{R}^n$, $g^{\dagger}(t,u,x) = x$, since on $T\times(\mathbb{R}^q-J_1)$, $m = 1$.

3) On $T\times\mathbb{R}^q\times(\mathbb{R}^n-B_1)$, $g^{\dagger}(t,u,x) = Q^{\dagger}(t,u,x)x$.

To complete the proof we construct $Q^* : T\times\mathbb{R}^q\times\mathbb{R}^n \longrightarrow R$ such that

$$T\times\mathbb{R}^q\times\mathbb{R}^n \ni (t,u,x) \longrightarrow (t,u,Q^*(t,u,x)x) \in T\times\mathbb{R}^q\times\mathbb{R}^n$$

is a diffeomorphism and

1') On $T\times\mathbb{R}^q\times B_0'$, $Q^* = Q^{\dagger}$.

2') On $T\times(\mathbb{R}^q\times\mathbb{R}^n-J_1\times B_1')$, $Q^* = I$.

Using the function $O$ obtained in the diagonalization of $Q'$, we define $Q^* = O\delta^*O$ where $\delta^* : T\times\mathbb{R}^q\times\mathbb{R}^n \longrightarrow D$ is an appropriate modification of $\delta^{\dagger}$ Let $r_0 < r_1$ be the radii of $B_0'$ and $B_1'$ respectively and let $\vartheta : \mathbb{R} \longrightarrow [0,1]$ be a smooth function such that $\vartheta' \geqslant 0$ everywhere and $\vartheta \mid (-\infty,r_0^2] = 0$ and $\vartheta \mid [r_1^2,\infty) = 1$. Define $\delta^* : T\times\mathbb{R}^q\times\mathbb{R}^n \longrightarrow D$ so that the diagonal entries of $\delta^*$ are:

$$d_i^*(t,u,x) = d_i^{\dagger}(t,u) + \vartheta(|x|^2)(1-d_i^{\dagger}(t,u)).$$

On $T\times(\mathbb{R}^q-J_1)\times\mathbb{R}^n$, $\delta^* = \delta^{\dagger} = I$, on $T\times\mathbb{R}^q\times B_0'$, $\delta^* = \delta^{\dagger}$, and on $T\times\mathbb{R}^q\times(\mathbb{R}^n-B_1')$, $\delta^* = I$. We check that for each $(t,u) \in T\times\mathbb{R}^q$, $x \longrightarrow \delta^*(t,u,x)x$ is a diffeomorphism of $\mathbb{R}^n$. To see that it is 1:1, suppose $\delta^*(t,u,x)x = \delta^*(t,u,x')x'$. If $|x| = |x'|$, then $\delta^*(t,u,x) = \delta^*(t,u,x')$ so $x = x'$. If $|x| > |x'|$, then for some $i$, $|x_i| > |x_i'|$. But $\vartheta$ is non-decreasing so $d_i^*(t,u,x) \geqslant d_i^*(t,u,x') > 0$. Thus $|d_i^*(t,u,x)x_i| > |d_i^*(t,u,x')x_i'|$ contradiction.

To compute the Jacobian of this map notice that its $i^{th}$ row is $d_i^*e_i + (\vartheta(|x|^2))_{x_i}(I-\delta^{\dagger})x$, where $e_i$ is the $i^{th}$ standard basis element of $\mathbb{R}^n$. Letting $\vartheta(|x|^2)_{x_i} = a_i$ and $(I-\delta^{\dagger})x = v$ and $d_i^*e_i = w_i$, $\bigwedge_i (w_i+a_iv) = \bigwedge_i w_i + \sum_{i=1}^n a_iw_i\wedge\cdots\wedge w_{i-1}\wedge v\wedge w_{i+1}\wedge\cdots\wedge w_n$. Thus the determinant of the Jacobian matrix is

$$(\prod_i d_i^*) + \sum_{i=1}^n \vartheta'(|x|^2)(d_1^*\cdots d_{i-1}^*(1-d_i^{\dagger})d_{i+1}^*\cdots d_n^*)x_i^2$$

which is positive.

Thus defining $Q^* = O\delta^* O^T$, the map $(t,u,x) \longrightarrow (t,u,Q^*(t,u,x)x)$ is a diffeomorphism as required.

Finally we define $g : T \times \mathbb{R}^q \times \mathbb{R}^n \longrightarrow \mathbb{R}^n$ by:

$$g(t,u,x) = \begin{cases} g^\dagger(t,u,x), & (t,u,x) \in T \times \mathbb{R}^q \times B_0' \\ Q^*(t,u,x)x, & (t,u,x) \in T \times \mathbb{R}^q \times \mathbb{R}^n - B_1 \end{cases}$$

This makes sense since for $x \in B_0' - B_1$, $g^\dagger(t,u,x) = Q^\dagger(t,u,x)$ and for $x \in B_0'$, $Q^\dagger(t,u,x) = Q^*(t,u,x)$. Thus $g$ is smooth and for each $(t,u)$, $g(t,u,\cdot)$ is an immersion of $\mathbb{R}^n$ into itself. However since $g^\dagger(t,u,\cdot)$ and $\Omega^*(t,u,\cdot)$ are both diffeomorphisms, which agree on $B_0' - B_1$, we know that $g^\dagger(t,u,\cdot)(B_0') = Q^*(t,u,\cdot)(B_0')$ and $g^\dagger(t,u,\cdot)(\mathbb{R}^n - B_1) = Q^*(t,u,\cdot)(\mathbb{R}^n - B_1)$. Thus $g(t,u,\cdot)$ is 1:1 as well, so $1 \times g$ is a diffeomorphism. Obviously

1') On $T \times \mathbb{R}^q \times B_0$, $g(t,u,x) = g^\dagger(t,u,x) = Q(t,u)x$.

2') On $T \times (\mathbb{R}^q \times \mathbb{R}^n - J_1 \times B_1')$, $g(t,u,x) = Q^*(t,u,x)x = x$ as required. // b)

c) $A = m \cdot I$ where $m : T \times \mathbb{R}^q \longrightarrow [1,\infty)$ and $m \mid T \times (\mathbb{R}^q - J_1) = 1$. Choose any positive real number $\tau$, and let $\sigma = \tau/2m$. Let $\psi$ : $T \times \mathbb{R}^q \times \mathbb{R} \longrightarrow \mathbb{R}$ be a smooth function such that

(i) $$\psi(t,u,s) = \begin{cases} m^2(t,u)s & \text{if } s \leqslant \sigma^2(t,u) \\ s & \text{if } s \geqslant \tau^2. \end{cases}$$

(ii) $0 < s(\partial\psi/\partial s)(t,u,s) \leqslant \psi(t,u,s)$, (we will write $\psi'$ for $\partial\psi/\partial s$).

To construct such a function choose $\sigma'$ and $\theta'$ so that $\sigma < \sigma' < \tau' < \tau$ for all $(t,u) \in T \times \mathbb{R}^q$ and so that $m\sigma' < \tau'$. Let $P(t,u,s)$ be the function, piecewise linear in $s$, given by:

$$P(t,u,s) = \begin{cases} m^2 s, & s \leqslant \sigma'^2 \\ \left[\dfrac{\tau'^2 - m^2\sigma'^2}{\tau'^2 - \sigma'^2}\right](s-\sigma'^2) + m^2\sigma'^2, & \sigma'^2 \leqslant s \leqslant \tau'^2 \\ s, & s \geqslant \tau'^2 \end{cases}$$

The map $\psi$ is obtained from $P$ by smoothing the corners at $s = \sigma'^2$ and $s = \tau'^2$.

Define $g : T \times \mathbb{R}^q \times \mathbb{R}^n \longrightarrow \mathbb{R}^n : (t,u,x) \longrightarrow \left(\dfrac{\psi(t,u,|x|^2)}{|x|^2}\right)^{1/2} x.$

Let $B_1$ be the ball of radius $\tau$ in $\mathbb{R}^n$ and let $N_0 = \{(t,u,x) \mid |x| \leq \sigma(t,u)\}$. For $(t,u,x) \in N_0$, $g(t,u,x) = m(t,u)x = A(t,u)\cdot x$. For $x \notin B_1$, $g(t,u,x) = x$. Finally for $u \in \mathbb{R}^q - J_1$, $m(t,u) = 1$, properties i) and ii) imply that $\psi(t,u,s) = s$ for all $s$, so $g(t,u,x) = x$. Thus to finish the proof we must check that $1 \times g$ is a diffeomorphism of $T \times \mathbb{R}^q \times \mathbb{R}^n$.

If $g(t,u,x) = g(t,u,x')$ then $\psi(t,u,|x|^2) = \psi(t,u,|x'|^2)$. But $\psi$ is monotonically increasing in the last variable so $|x| = |x'|$. Thus $x = x'$. Computing the $(ij)^{th}$ entry in the Jacobian of $g$: in the region $\sigma^2 \leq |x|^2 \leq \tau^2$.

$$\frac{\partial g_j}{\partial x_i} = \left(\frac{\psi}{|x|^2}\right)^{1/2}\left(\delta_{i,j} - \frac{\psi - |x|^2\psi'}{|x|^2\phi}\, x_i x_j\right).$$

Since we know that $\psi - |x|^2\psi' \geq 0$ by construction of $\psi$, we apply the following easy lemma:

Lemma. $\det(\delta_{ij} - y_i y_j) = (1-|y|^2)$, $1 \leq i,\ j \leq n$.

Thus the value of the Jacobian determinant is:

$$\left(\frac{\psi}{|x|^2}\right)^{n/2}\left(1 - \left(\frac{\psi - |x|^2\psi'}{\psi}\right)\right) = \left(\frac{\psi}{|x|^2}\right)^{n/2}\left(\frac{|x|^2\psi'}{\psi}\right) > 0. \qquad //$$

## A-3.4 Lemma

Lemma. *Let $1 \times k : I \times \mathbb{D} \longrightarrow I \times \mathbb{R}^2$ be an embedding with either*

1) $k(u,(0,0)) = 0$.

2) $k(3\varepsilon x^2,(x,0)) = (x,0)$.

*Then there is a regular homotopy* $1 \times k_t$ *with* $k_0 = k$ *and*

(i) $k_t$ *is independent of* $t$ *in a neighborhood of the boundary of* $I \times \mathbb{D}$.

(ii) *On a sufficiently small neighborhood* $(I' \times \mathbb{D}')$ *of* $(0,0)$, $k_1(u,x,y) = (x,y)$, *(or* $= (x,-y))$ *if* $1 \times k$ *is orientation preserving, (or reversing).*

(iii) *If* $V$ *is an open set in* $\mathbb{D}$ *and* $u_0 \in I$ *such that* $k(u_0,(x,y)) = (x,\pm y)$ *for* $(x,y) \in V$, *then* $k_t(u_0,(x,y)) = (x,\pm y)$ *for all* $t$ *and* $(x,y) \in V$.

(iv) *Under hypotheses:* 1) $k_t(u,(0,0)) = 0$, *and* 2) $k_t(3\varepsilon x^2,(x,0)) = (x,0)$ *for all* $t$.

Proof. When convenient we will consider $\mathbb{D} \subseteq \mathbb{C} = \mathbb{R}^2$ and denote $(x,y)$ by $z$ and $(x,-y)$ by $\bar{z}$. It suffices to prove the lemma for

$(1\times k)$ orientation preserving, for if $1 \times k$ reversed orientation apply the lemma to $1 \times \bar{k}$ to obtain $k_t'$. Our required homotopy is then $\bar{k}_t'$.

Expand $k(u,(x,y))$ about $(0,0)$ in $I \times \mathbb{D}$:

$$k(u,z) = xz + yb + \{k(u,0) + xu\ell(u) + yum(u) + R(u,z)\}$$

where $\|R(u,z)\|/\|z\| \longrightarrow 0$ as $z \longrightarrow 0$ for all $u \in I$, and $a, b \in \mathbb{R}^2$. We write $k(u,z) = A(x,y) + G(u,(x,y))$ where $A$ is a nonsingular linear transformation. Write $G(u,(x,y)) = H(u,A(x,y))$, so that $1 \times k$ is the composition of $1 \times A$ and $1 \times h$ where $h(u,(x,y)) = (x,y) + H(u,(x,y))$, and $H$ has no linear terms in $x$ or $y$. Under hypotheses 1), $k(u,0) = 0$, we have $G(u,0) = - = H(u,0) = h(u,0)$, and so $(1\times h)$ satisfies hypothesis 1). Under hypothesis 2),

$$k(3\varepsilon x^2,(x,0)) = (x,0) = A(x,0) + G(3\varepsilon x^2,(x,0)).$$

But all the terms in $G(3\varepsilon x^2,(x,0))$ have order larger than one at $(0,0)$ so we see that $A(x,0) = (x,0)$ and $G(3\varepsilon x^2(x,0)) = 0$. Hence $0 = G(3\varepsilon x^2,(x,0)) = h(3\varepsilon x^2,A(x,0)) = H(3\varepsilon x^2,x,0)$, and so $h(3\varepsilon x^2,(x,0)) = (x,0)$. Thus here again $1 \times h$ satisfies hypothesis 2).

Let $\phi : \mathbb{R} \longrightarrow [0,1]$ be a smooth function with $\operatorname{supp}\phi = [1,\infty)$, $\phi \mid [2,\infty) = 1$ and $0 \leqslant \phi' < 2$. For $x \in \mathbb{R}^n$, let $\psi_r(x) = \phi(\|x\|^2/r^2)$ for any $r > 0$. $|(\partial\psi_r/\partial x_i)(x)| = \phi'(\|x\|^2/r^2)(2|x_i|/r^2) < 6/r$. On $\mathbb{R} \times \mathbb{R}^n$, let $\vartheta_{r,s}(u,x) = \psi_s(u) + \psi_r(x) - \psi_s(u)\cdot\psi_r(x)$. Clearly $\vartheta_{r,s}(u,x) = 0$ if $|u| \leqslant s$ and $\|x\| \leqslant r$ and $\vartheta_{r,s}(u,x) = 1$ if $|u| \geqslant \sqrt{2}s$ or $\|x\| \geqslant \sqrt{2}r$. Finally $|\partial\vartheta_{r,s}/\partial x_i(u,x)| < 6/r$.

We consider first the case $k(u,(x,y)) = (x,y) + G(u,(x,y))$, where $G(u,(x,y)) = k(u,0) + ux\ell(u) + uym(u) + R(u,(x,y))$ and $\|R(u,(x,y))\|/\|(x,y)\| \longrightarrow 0$ as $\|(x,y)\| \longrightarrow 0$. Under hypotheses 1) and 2), $G(u,0) = 0$ and $G(3\varepsilon x^2,(x,0)) = 0$ respectively. Let $k_t(u,(x,y)) = (x,y) + G(u,(x,y))(1-t+t\vartheta_{r,s}(u,(x,y)))$. This homotopy satisfies conclusions i) through iv). We must check that for appropriately chosen $r$ and $s$, that $1 \times k_t$ is an immersion for all $t \in [0,1]$. We compute the Jacobian of $k_t$ with respect to $x$ and $y$ as:

$$(\partial k_t/\partial x) \wedge (\partial k_t/\partial y)$$

$$= [(1,0) + ((\partial k_t/\partial x)-(1,0))] \wedge [(0,1) + ((\partial k_t/\partial y)-(0,1))].$$

It is enough to show that we can choose $r$, $s$ small enough so that $\|(\partial k_t/\partial x)-(1,0)\|$ and $\|(\partial k_t/\partial y)-(0,1)\|$ are smaller than $\varepsilon$ for sufficiently small $\varepsilon$--say $\varepsilon < 1/3$, if $|u| \leqslant \sqrt{2}s$ and $\|(x,y)\| \leqslant \sqrt{2}r$.

$$\|(\partial k_t/\partial x)(u,z)-(1,0)\|$$

$$= \|(\partial G/\partial x)(u,z)\{1-t+t\vartheta_{r,s}(u,z)\}+G(u,z)t(\partial\vartheta_{r,s}(u,z)/\partial x)\|$$

$$< \|(\partial G/\partial x)(u,z)\| + \|G(u,z)\|(6/r)$$

$$< \|u\ell(u)\| + \|(\partial R/\partial x)(u,z)\|$$

$$+ (6/r)(\|k(u,0)\|+r(\|u\ell(u)\|+\|um(u)\|+\|R(u,z)\|).$$

Since $(\partial R/\partial x)(u,0) = 0$ and $\|R(u,z)\|/\|z\| \longrightarrow 0$ as $\|z\| \longrightarrow 0$ we can choose $r$ and $s_1$ so that if $\|z\| \leqslant \sqrt{2}r$ and $|u| \leqslant s_1$, then $\|(\partial R/\partial x)(u,z)\| \leqslant \varepsilon/3$ and $\|R(u,z)\|/r \leqslant (\sqrt{2}/\|z\|)\|R(u,z)\| < \varepsilon/18$. For such a choice of $r$ and $s_1$, if $\|z\| \leqslant \sqrt{2}r$ and $|u| < s_1$

$$\|(\partial k_t/\partial x)(u,z)-(1,0)\|$$

$$< (6/r)\|k(u,0)\| + 7\|u\ell(u)\| + 6\|um(u)\| + (2\varepsilon/3)$$

since $k(0,0) = 0$, we can choose an $s > 0$ so small that if $|u| \leqslant \sqrt{2}s < s_1$, then the sum of the first three terms is bounded by $\varepsilon/3$. In a completely analogous way we can make $\|(\partial k_t/\partial y)-(0,1)\| < \varepsilon$. Thus for appropriately chosen $(r,s)$ we see that $1 \times k_t$ is the required regular homotopy joining $1 \times k$ to a map $1 \times k_1$ which is the identity on $|u| \leqslant s$, $\|z\| \leqslant r$.

We now treat the case: $k(u,x,y) = A(x,y)$, $A$ linear, non-singular. Since $k(u,0,0) = A(0,0) = 0$ is automatic for $A$ linear we treat that case first. Factor $A$ as $\mathcal{O}_1\Delta\mathcal{O}_2$ where $\mathcal{O}_i \in SO(2)$ and $\Delta(x,y) = (ax,by)$ for some $a > 0$, $b > 0$. We write $1 \times A$ as $(1\times\mathcal{O}_1)\circ(1\times\mathcal{O})\circ(1\times\mathcal{O}_2)$. Thus we need merely prove the lemma for $1 \times A$ where $A \in SO(2)$ and where $A$ is positive diagonal. If $A \in SO(2)$, let $\Phi : [0,1] \longrightarrow SO(2)$ be a smooth map with $\Phi_0 = A$ and $\Phi_1 = I$. Let $\eta : I \times \mathbb{D} \longrightarrow [0,1]$ be a smooth function that vanishes in a neighborhood of the boundary of $I \times \mathbb{D}$ and has value $1$ on a neighborhood of $(0,0) \in I \times \mathbb{D}$. Further we suppose that $\eta(u,x,y)$ depends only on $u^2$ and $x^2 + y^2$. Define $A_t(u,x,y) = \Phi_{t\eta(u,x,y)}(x,y)$. Obviously $A_0(u,x,y) = A(x,y)$.

On the neighborhood of the boundary of $I \times \mathbb{D}$ where $\eta$ vanishes, $A_t = A$, for all $t$. Finally in the neighborhood of $0$ where $\eta = 1$, $A_1(u,x,y) = (x,y)$. We check that $(u,x,y) \longrightarrow u, A_t(u,x,y)$ is a diffeomorphism. Claim: $(u,x,y) \longrightarrow \Phi^T_{t\eta(u,x,y)}(x,y) = (u,B_t(u,x,t))$ is the inverse of $(1 \times A_t)$. Composing the maps gives

$$\begin{aligned}(u,B_t(u,A_t(u,x,y))) &= (u,\Phi^T_{t\eta(u,A_t(u,x,y))}\cdot A_t(u,x,y)) \\ &= (u,\Phi^T_{t\eta(u,\Phi_t\ (u,x,y)}(x,y))\cdot\Phi_{t\eta(u,x,y)}(x,y)) \\ &= (u,\Phi^T_{t\eta(u,x,y)}\cdot\Phi_{t\eta(u,x,y)}(x,y)) \\ &= (u,(x,y)),\end{aligned}$$

since $\|\Phi_{t\eta(u,x,y)}(x,y)\|^2 = \|(x,y)\|^2$, so $\eta(u,\Phi_{t\eta(u,x,y)}(x,y)) = \eta(u,x,y)$.

If $A(x,y) = (ax,by)$, let $\alpha : \mathbb{R} \longrightarrow \mathbb{R}$ be a smooth function such that $\alpha(x) = x$ for $|x| \leqslant r_0$ and $\alpha(x) = ax$ for $|x| \geqslant r_1$ and $\alpha' > 0$ everywhere for any $r_1 > (1/a)r_0 > 0$. Let $\rho : \mathbb{R} \times \mathbb{R} \longrightarrow [0,1]$ be a smooth function such that $\rho(u,y) = 1$ on $(u^2+y^2) \leqslant s_0^2$ and $\rho(u,y) = 0$ on $(u^2+y^2) \geqslant s_1^2$ for $s_1 > s_0 > 0$.

Let $A_t(u,(x,y)) = (\alpha(x)t\rho(u,y) + ax(1-t\rho(u,y)),by)$. Obviously for all $t \in [0,1]$, $1 \times A_t$ is a diffeomorphism. If $|x| \leqslant r_0$ and $(u^2+y^2) \leqslant s_0^2$, $A_1(u,(x,y)) = (x,by)$, and if $|x| \geqslant r_1$ or $(u^2+y^2) \geqslant s_1^2$, $A_t(u,(x,y)) = A(u,(x,y))$. Now inside the neighborhood of $(0,0)$ in which $A_1(u,(x,y)) = (x,by)$, deform $(1 \times A_1)$ in an analogous manner, interchanging the roles of $x$ and $y$. This completes the proof of the case $k(u,(x,y)) = A(x,y)$, $A$ linear and $k(u,(0,0)) = 0$.

In case $k(u,(x,y)) = A(x,y)$, $A$ linear and nonsingular but $k(3\varepsilon x^2,x,0) = A(x,0) = (x,0)$ we know that $A(x,y) = (x+ay,by)$, where $b > 0$ since $1 \times A$ preserves orientation. We deform this to the diagonal case. By a slight variation of what we did at the beginning of the proof of the proposition, given any positive real numbers $(r,s)$ we can construct a function $\eta : \mathbb{R}^3 \longrightarrow [0,1]$ such that if $q = r/9a$, then $\eta(u,x,y) = 1$ if $|u| \leqslant s$, $|x| \leqslant r$, and $|y| \leqslant q$; $\eta(u,x,y) = 0$ if $|u|^2 \geqslant 2s^2$ or $|x|^2 \geqslant 2r^2$ or $|y|^2 \geqslant 2q^2$ and $|\partial\eta/\partial x| < 6/r$ everywhere.

Define:

$$A_t(u,x,y) = (x+ay(1-t\eta(u,x,y)),by).$$

Notice that $A_t(u,x,0) = (x,0)$ for all $u, t$; $A_0 = A$, if $u^2 \geq 2s^2$, $|x|^2 \geq 2r^2$ or $|y|^2 \geq 2q^2$, $A_t = A$. If $|u| \leq s$, $|x| \leq r$ and $|y| \leq q$, $A_t(u,x,y) = (x+(1-t)ay, by)$, so in that set $A_1(u,x,y) = (x,by)$.

We check that $1 \times A_t$ is a diffeomorphism for all $t$. To check that $(1 \times A_t)$ is 1:1 we must show that if

$$(1) \qquad x + ay(1-t\eta(u,x,y)) = x' + ay(1-t\eta(u,x',y))$$

then $x = x'$. It suffices to show this if $|u|^2 \leq 2s^2$ and $|y|^2 \leq 2q^2$, since it is automatic otherwise. If $x > x'$, we have

$$(1') \qquad 1 = tay\left(\frac{\eta(u,x,y)-\eta(u,x',y)}{x - x'}\right) = tay\left(\frac{\partial\eta(u,\bar{x},y)}{\partial x}\right)$$

for some $\bar{x} \in (x',x)$. But $|tay(\partial\eta/\partial x)(u,\bar{x},y)| < (a\sqrt{2}q6/r) < (2\sqrt{2}/3) < 1$, contradiction. Thus $1 \times A_t$ is 1:1.

Similarly the non-singularity of the Jacobian of $1 \times A_t$ requires that for $|u|^2 \leq 2s^2$, $|x|^2 \leq 2r^2$, $|y|^2 \leq 2q^2$, $1 - tay(\partial\eta/\partial x)(u,x,y) \neq 0$ which we have just checked. Thus we've reduced this case to that of $A(x,y) = (x,by)$, $b > 0$ which we have already dealt with. To see that the previous proof applies to this case we must check that if $A_t$ is the deformation used there, that $A_t(3\varepsilon x^2,x,0) = (x,0)$. This is obvious since the deformation has the form: $A_t(u,(x,y)) = (x, \beta(y)t\rho(u,x) + by(1-t\rho(u,x)))$, where $\beta(y) = y$ for $y$ in a neighborhood of $0$. //

## A-3.6 Proposition 2 and Proposition 3

Proposition 2. Let $(h_1)$, $(h_0) \in H_{S,\Sigma}$. The following are equivalent:

1) $\lambda_{h_0} = \lambda_{h_1}$.

2) If $h_0$ and $h_1$ are any representatives of $(h_0)$ and $(h_1)$ with a common domain $B(\Sigma)$, then there is a smooth homotopy $H : h_0 = H_0 \simeq H_1$ such that for all $t \in [0,1]$

(i) $(f,H_t)$ immerses $B(\Sigma)$ in $\mathbb{R}^4$.

(ii) $H_t$ is independent of $t$ in a neighborhood of $S(f)$ and on a neighborhood of $\partial B(\Sigma)$.

(iii) $(H_1)_{\hat{\Sigma}} = (h_1)_{\hat{\Sigma}}$.

Proof. If $(h_0)$ and $(h_1)$ satisfy 2) then trivially $\delta_{h_0} = \delta_{h_1}$, and $\tau_{h_0} = \tau_{h_1}$. We show that $\sigma_{h_0} = \sigma_{h_1}$. Let $\alpha \in A$. Since $h_0 = H_0$ and

$(h_1)_{\hat{\Sigma}} = (H_1)_{\hat{\Sigma}}$ it is enough to show that $\mathrm{rot}(H_t \circ \bar{\alpha})$ is independent of $t$. Since $\bar{\alpha}(0)$ and $\bar{\alpha}(1) \in S(f)$ and $H_t$ is independent of $t$ in a neighborhood of $S(f)$, the map $\Lambda = H \circ \bar{\alpha} : I \times I \longrightarrow \mathbb{R}^2 : t, s \longrightarrow H_t(\bar{\alpha}(s))$ is independent of $t$ on a neighborhood of $(I \times \partial I)$. Thus $(\partial\Lambda/\partial s)(t,0)$ and $(\partial\Lambda/\partial s)(t,1)$ are constants, and so defining $\theta : I \times I \longrightarrow \mathbb{R}$ by $\theta(t,0) \in [0,1)$ and

$$(\partial\Lambda/\partial s)(t,s) = \|(\partial\Lambda/\partial s)(t,s)\| e^{2\pi i \theta(t,s)}$$

we see that $\theta(t,0)$ and $\theta(t,1)$ are constant so $\theta(t,1) - \theta(t,0) = \mathrm{rot}(H_t \circ \bar{\alpha})$ is independent of $t$.

Now suppose $\lambda_{h_0} = \lambda_{h_1}$, and $B(\Sigma)$ a neighborhood of $\hat{\Sigma}$ is in the domain of both $h_0$ and $h_1$. Since $(h_0)_{S(f)} = (h_1)_{S(f)}$ we may assume that $h$ and $h'$ agree on a tubular neighborhood of $S(f)$. For all $t \in [0,1]$, $H_t$, $h_0$ and $h_1$ will coincide in a slightly smaller tubular neighborhood of $S(f)$. Our homotopy will be constructed in two stages. In the first we will alter $h_0$ only in a neighborhood of $q^{-1}(v)$ for all vertices $v \in V$, and in a neighborhood of $q^{-1}(v_c)$ for $v_c$ a point on a closed component $c \in C$. In the second stage we complete the homotopy by further changing the resulting $h$ near $q^{-1}(c)$ for $c \in C$.

If $c \in C_0$, since $q^{-1}(c)$ is just an open arc of $S(f)$, a neighborhood of $q^{-1}(c)$ is contained in the neighborhood of $S(f)$ on which $h_0 = h_1 = h_{\delta,\tau}$. Thus we will work only in neighborhoods of $q^{-1}(v)$ for $v \in V$, and $q^{-1}(v_c)$ for $v_c \in c$, $c$ a closed component of $C_1$, and then in neighborhoods of $q^{-1}(c)$ for $c \in C_1$.

Both parts of the homotopy will follow from an easy generalization of the Whitney-Graustein Theorem classifying immersions of circles in the plane under regular homotopy [W-G].

We begin with a $C^2$PN for $f$ at $v \in V$.

$$\begin{array}{ccc} I \times T(v) & \xrightarrow{\phi_v} & B(v) \\ {\scriptstyle 1 \times g_v}\downarrow & & \downarrow{\scriptstyle f} \\ I \times J & \xrightarrow{\psi_v} & \mathbb{R}^2 \end{array}$$

where we assume $B(v) \subseteq B(\Sigma)$. At this stage, we will not change $h$ outside of a neighborhood of $q^{-1}(v)$ in $B(v)$, so we work completely inside $I \times T(v)$ with maps: $(k_0)_v = h_0 \circ \phi_v$ and $(k_1)_v = h_1 \circ \phi_v$. In case $c \in C_1$, $c$ closed, we will work completely inside $I_- \times T(c)$

(see §2.3) with the maps $(k_0)_{c,-} = h_0 \circ \phi_{c,-}$ and $(k_1)_{c,-} = h_1 \circ \phi_{c,-}$ on $I_- \times T(c)$.

In the hope of simplifying the notation without diminishing the intelligibility we suppress, v, c, +, - as much as possible.

Thus we have three maps, $k_0, k_1 : I \times T \longrightarrow \mathbb{R}^2$ and $1 \times g : I \times T \longrightarrow I \times J$ such that $1 \times g \times k_0$ and $1 \times g \times k_1$ are immersions. In case c is a closed component, there is a disc $\mathbb{D}$ about 0 in T(c), if v is a cusp, there is a disc $\mathbb{D}$ about 0 in T(v), and if v is a nonsimple vertex there are two discs $\mathbb{D}_\pm$ about points $(\pm x_0, 0)$ in T(v) such that on $I \times \mathbb{D}$ or $I \times \mathbb{D}_\pm$, $k_0$ and $k_1$ coincide and such that for each $u \in I$, $k_0$ and $k_1$ embed $u \times \mathbb{D}$ or $u \times \mathbb{D}_\pm$.

For each of these discs we let $\mathbb{D}' \subseteq \mathbb{D}$ be a smaller concentric disc such that $I \times \mathbb{D}'$ (or $I \times \mathbb{D}_\pm'$) contains $S(1\times g) = \phi^{-1}(S(f))$. In each of these cases for some open interval $J' \subseteq J$, there are subsets of $I \times T$ diffeomorphic to $I \times J' \times \mathbb{S}^1$ the union of which together with the above mentioned disc neighborhoods $I \times \mathbb{D}'$ contains $\phi^{-1}(\hat{\Sigma}) = (1\times g)^{-1}(1\times g)(S(1\times g))$. Further we can take the diffeomorphism $d : I \times J' \times \mathbb{S}^1 \longrightarrow I \times T$ so that $d^{-1}(I\times\mathbb{D}') \subseteq d^{-1}(I\times\mathbb{D})$ (or $d^{-1}(I\times\mathbb{D}_\pm') \subseteq d^{-1}(I\times\mathbb{D}_\pm)$) has the form $I \times J' \times U' \subseteq I \times J' \times U$ where $U' \subseteq U$ are open arcs on $\mathbb{S}^1$. In the cases in which we delete $d^{-1}(I\times\mathbb{D}_\pm') \subseteq d^{-1}(I\times\mathbb{D}_\pm)$ (v of type (1·2·2·1) or (1·2·2·3)), $U' \subseteq U$ will consist of four open arcs. Finally we can choose these diffeomorphisms so that $d\circ(1\times g) : I \times J' \times (\mathbb{S}^1 - U) \longrightarrow I \times J$ is just projection, i.e. the d-images of $(u\times v\times(\mathbb{S}^1-U))$ are arcs of the fibres of $1 \times g$. We picture these sets in the transverse manifolds in the four cases.

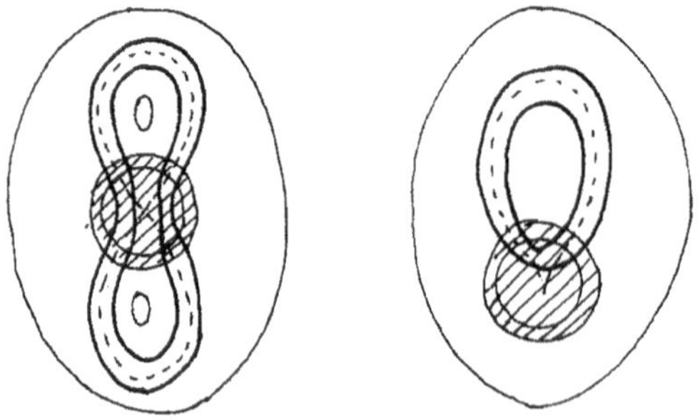

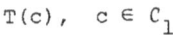

T(c), $c \in C_1$ T(v), v a cusp

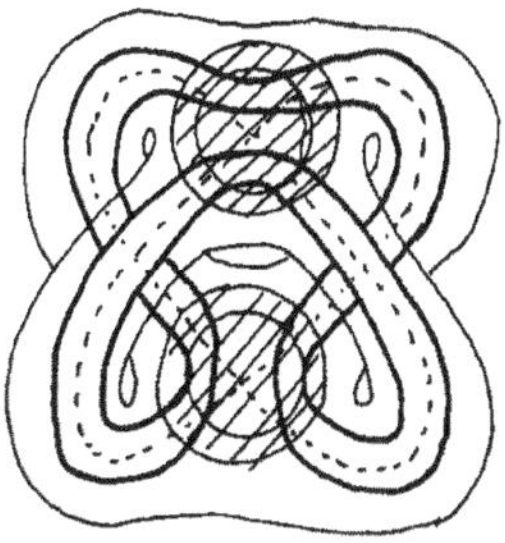

$T(v)$, $v$ a $(1\cdot 2\cdot 2\cdot 1)$ vertex

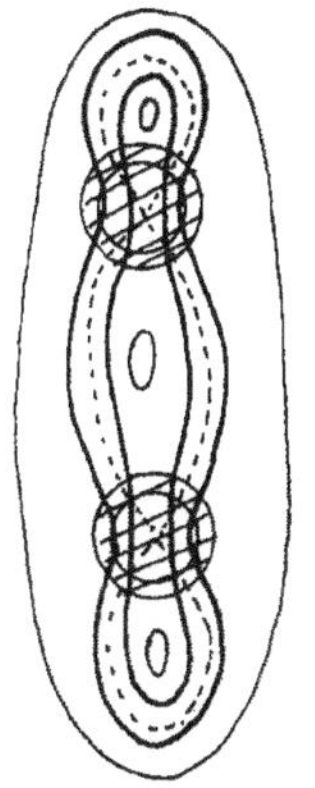

$T(v)$, $v$ a $(1\cdot 2\cdot 2\cdot 3)$ vertex

In these figures the discs $\mathbb{D}' \subseteq \mathbb{D}$ and $\mathbb{D}'_{\pm} \subseteq \mathbb{D}_{\pm}$ are shaded. The d-images of $J' \times \mathbb{S}^1$ are heavily outlined. The dotted curves are the $(1\times g)$-fibres through the indicated singular points of $(1\times g)$. We now suppress the embeddings, d, from our notation and drop the primes in the $J'$.

Thus we have maps, $k_0, k_1 : I \times J \times \mathbb{S}^1 \longrightarrow \mathbb{R}^2$, where $I$ and $J$ are open intervals. We translate the hypotheses of our proposition that we will use for this stage of the homotopy to these maps.

1) $k_i \mid u \times v \times \mathbb{S}^1$ are immersions for all $u \in I$, $v \in J$, $i = 0, 1$.

2) For a finite union of disjoint open arcs of $\mathbb{S}^1$, $U$, $k_0 \mid u \times J \times U = k_1 \mid u \times J \times U$ are embeddings for all $u \in I$.

3) Orient $\mathbb{S}^1$, then on each arc $\alpha$ of $u \times v \times (\mathbb{S}^1 - U')$ $\mathrm{rot}(k_0 \circ \alpha) = \mathrm{rot}(k_1 \circ \alpha)$.

The hypothesis 1) and the part of 2) that $k_i \mid u \times J \times U$ are embeddings are a consequence of $(h_0)$ and $(h_1) \in H_{S,\Sigma}$. The equality $k_0 \mid u \times J \times U = k_1 \mid u \times J \times U$ results from the additional hypothesis, $\delta_{h_0} = \delta_{h_1}$. Finally 3) is a consequence of the equality $\sigma_{h_0} = \sigma_{h_1}$ (i.e. $\nu_{h_0} \mid A = \nu_{h_1} \mid A$). The first stage of our homotopy is obtained by applying the following lemma to $k_0, k_1$.

*Lemma 1. Let* $k_0, k_1 : I \times J \times \mathbb{S}^1 \longrightarrow \mathbb{R}^2$ *satisfy 1), 2), 3) above. Let* $I' \subseteq I$ *and* $J' \subseteq J$ *be nontrivial closed subintervals. Then there is a homotopy* $K : k_0 = K_0 \simeq K_1$ *such that for all* $t \in [0,1]$ $K_t$ *satisfy:*

(i) $K_t \mid u \times v \times \mathbb{S}^1$ *is an immersion.*

(ii) $K_t \mid I \times J \times U = k_0 \mid I \times J \times U$.

(iii) $K_t = k_0$ on a neighborhood of $\partial(I\times J) \times \$^1$.

(iv) $K_1 \mid I' \times J' \times \$^1 = k_1 \mid I' \times J' \times \$^1$.

This lemma, once proved, reduces the proof of the Proposition to the special case in which $h_0$ and $h_1$ coincide on a neighborhood of $q^{-1}(v)$, $B(v) \subseteq B(\Sigma)$ for each vertex $v \in V$ and on a neighborhood, $B(v_c)$ of $q^{-1}(v_c)$ for $v_c \in c$, where $c$ is a closed component of $_1$. By the earlier remark we know that $h_0$ and $h_1$ coincide on a neighborhood of each component $c \in C_0$. Now precisely as above we consider $h$ and $h'$ on a neighborhood $B(c)$ of $q^{-1}(c)$ for each $c \in C_1$. Let $(\phi,\psi,g)$ be a $C^2$PN for $f$ along $c$. We restrict our attention to $\phi(I\times T) \in B(c) \subseteq B(\Sigma)$, and let $k_0 = h_0\circ\phi$ and $k_1 = h_1\circ\phi$.

Again we have an open subset $I \times \mathbb{D} \subseteq I \times T$ on which $k_0$ and $k_1$ coincide and subsets $(I\times J\times \$^1) \subseteq I \times T$ such that $(I\times\mathbb{D}) \cup (I\times J\times\$^1)$ is a neighborhood of $\phi^{-1}(\hat{\Sigma})$. Our maps $k_0$ and $k_1$ on $(I\times J\times\$^1)$ satisfy the hypotheses 1), 2), 3) of Lemma 1). Now the equality in hypothesis 2) results from the additional equality, $\tau_{h_0} = \tau_{h_1}$. In addition we have $k_0 = k_1$ on $(I_\pm\times J\times\$^1)$ where $I_+$ and $I_-$ are non-empty open intervals about the ends of $I$ such that $\phi(I_\pm\times J\times\$^1)$ are inside the sets, $B(v)$ or $B(v_c)$ on which $h_0$ and $h_1$ agree. Thus to finish the proof of the proposition it suffices to prove:

<u>Lemma 1'</u>. <u>Let</u> $k_0, k_1 : I \times J \times \$^1 \longrightarrow \mathbb{R}^2$ <u>satisfy hypotheses 1), 2), 3) of Lemma 1 and in addition suppose</u>:

4) <u>For some nontrivial relatively closed intervals</u> $I'' \subseteq (I_1'')^\circ \subseteq I_1'' \subseteq I$, $k_0 = k_1$ <u>on</u> $(I-I'') \times J \times \$^1$. <u>Let</u> $I' \subseteq I$ <u>and</u> $J' \subseteq J$ <u>be nontrivial closed subintervals, then there is a homotopy</u> $L : k_0 = L_0 \simeq L_1$ <u>such that all</u> $t \in [0,1]$:

i) $L_t \mid u \times v \times \$^1$ <u>is an immersion</u>.

ii) $L_t = k_0$ <u>on</u> $I \times J \times U$.

iii) $L_t = k_0$ <u>on a neighborhood of</u> $\partial(I\times J) \times \$^1$.

iv) $L_1 = k_1$ <u>on</u> $(I' \cup (I-I_1'')) \times J' \times \$^1$.

Lemma 1 is the special case of Lemma 1' where $I'' = I$. If $I''$ were a closed subinterval of $I$ and we chose $I' \supseteq I''$ then conclusion iv) becomes $L_1 = k_1$ on $I \times J' \times \$^1$ which is the case we need for Proposition 1. We show that a slightly simpler version of Lemma 1' implies it.

Lemma 1". Let $I$ be an open interval and $J$ a closed interval. Let $k_0, k_1 : I \times J \times S^1 \longrightarrow \mathbb{R}^2$ be smooth maps such that:

1) For each $(u,v) \in I \times J$, $k_i \mid u \times v \times S^1$ are immersions, $i = 0, 1$.

2) For $U$ the union of $n$ disjoint open arcs in $S^1$ and $I" \subseteq I$ a nontrivial relatively closed interval, $k_0 = k_1$ on $(I-I") \times J \times S^1 \cup (I\times J\times U)$.

3) For each component arc $\alpha$ of $S^1 - U$, $\mathrm{rot}(k_0\circ\alpha) = \mathrm{rot}(k_1\circ\alpha)$.

If $I" \neq I$, let $I_1"$ be a closed interval such that $I" \subseteq (I_1")^\circ$. Then there is a smooth homotopy $M : k_0 \simeq k_1$ such that for all $t \in [0,1]$:

i) $M_t \mid (u\times v\times S^1)$ is an immersion.

ii) $M_t = k_0$ on $((I-I_1")\times J\times S^1) \cup (I\times J\times U)$.

Proof that 1" implies 1': Choose $I' \subseteq I$ and $J' \subseteq \tilde{J} \subset J$ any nontrivial closed subintervals. Let $e : [0,1] \times S^1 \longrightarrow (I\times\tilde{J}-I'\times J')$ be an embedding of the annulus so that deleting $e([0,1]\times S^1)$ from $I \times J$ leaves an open set $P \supseteq \partial(I\times J)$ and another $Q \supseteq I' \times J'$ and $\partial Q = e(1\times S^1)$. Let $\rho : [0,1] \longrightarrow [0,1]$ be smooth with $\rho \mid [0,1/3] = 0$ and $\rho \mid [2/3,1] = 1$ and $\rho' \geq 0$ everywhere.

Given $M$ as in Lemma 1", applied to $k_0, k_1$ restricted to $I \times J \times S^1$, we define $L$ as follows:

$$L(t,(u,v,\vartheta)) = \begin{cases} k_0(u,v,\vartheta), & \text{for } u,\ v \in P \\ M(t,(u,v,\vartheta)), & \text{if } (u,v) \in Q \\ M(t\rho(s),(u,v,\vartheta)), & \text{if } (u,v) = e(s,\sigma) \in I \times J - (P\cup Q) \end{cases}$$

Obviously $L$ satisfies condition i).

Since $P$ is a neighborhood of $\partial(I\times J)$ we have conclusion (iii) of Lemma 1' for L. Since $I' \times J' \subseteq Q$ and since $M_1 = k_1$, we have $L_1 = k_1$ on $I' \times J' \times S^1$. The rest of conclusion iv) as well as conclusion ii) of Lemma 1' follow from condition ii) of Lemma 1". // $(1" \Rightarrow 1')$

We make another minor reduction. If we denote by Lemma$_n$, Lemma 1" with $U$ having $n$ components, we show that Lemma$_1$ implies Lemma$_n$. We proceed by induction on $n$. Suppose we know Lemma$_1$ and Lemma$_{n-1}$. Let $k_0, k_1, I, J, I", U$ satisfy the hypotheses of Lemma$_n$ with $U = U_1 \cup \ldots \cup U_n$, $\{U_i\}$ disjoint open arcs of $S^1$. Suppose that $A_1,\ldots,A_n$ are the complementary arcs and that going around $S^1$ we encounter $U_1,A_1,\ldots,U_n,A_n$ in that order. Apply Lemma$_{n-1}$ to $\tilde{k}_0 = k_0$ and $\tilde{k}_1$, $I$, $J$, $I"$, $\tilde{U}$, where

$$\tilde{k}_1 = \begin{cases} k_1 & \text{on } I \times J \times (\$^1 - A_{n-1}) \\ k_0 & \text{on } I \times J \times A_{n-1} \end{cases} \quad \text{and} \quad \begin{cases} \tilde{U}_i = U_i, & i = 1,\ldots,n-2 \\ \tilde{U}_{n-1} = U_{n-1} \cup A_{n-1} \cup U_n \\ \tilde{U} = \bigcup_{i=1}^{n-1} \tilde{U}_i \end{cases}$$

We obtain a homotopy $\tilde{M} : k_0 = \tilde{k}_0 \simeq \tilde{k}_1$ where $\tilde{M}_t \mid u \times v \times \$^1$ is an immersion and $\tilde{M}_t = k_0$ on $(I-I'') \times J \times \$^1 \cup (I \times J \times \tilde{U})$. Now applying Lemma$_1$ to $\hat{k}_0 = \tilde{k}_1$ and $\hat{k}_1 = k_1$, $I$, $J$, $I''$, $\hat{U} = \$^1 - A_{n-1}$, we get a homotopy $\hat{M} : \tilde{k}_1 \simeq k_1$ with $\hat{M}_t \mid (u \times v \times S^1)$ an immersion and $\hat{M}_t = \tilde{k}_1$ on $(I-I'') \times J \times \$^1 \cup (I \times J \times \hat{U})$. Putting these homotopies together we get a homotopy which is constantly $k_0$ on $((I-I'') \times J \times \$^1) \cup (I \times J \times U)$. This constancy during $\tilde{M}$ is immediate since $U \subseteq \tilde{U}$. To see the constancy during $\hat{M}$, notice that $\hat{k}_1 = \tilde{k}_1 = k_1$ on $I \times J \times \hat{U}$, but $k_1 = k_0$ on $I \times J \times U$ and $\hat{k}_1 = \tilde{k}_1 = k_1 = k_0$ on $(I-I'') \times J \times \$^1$.

// (Lemma$_1$ ⇒ Lemma$_n$)

Thus it is sufficient to prove the case of Lemma 1" for which $U$ has a single component.

We now state and prove a proposition that is slightly more general than Lemma 1". It is an easy generalization of the Whitney-Graustein classification theorem for immersions of circles in the plane under regular homotopy [W-G] and the proof is theirs except for minor modifications because of the existence of parameters.

Let $X$ be a smooth manifold possibly with boundary which may have corners. Let $J = [0,1]$. A smooth map $g : X \times J \longrightarrow \mathbb{R}^2$ is called a <u>family of regular closed curves parametrized by</u> $X$ if

1) $(\partial g/\partial t)(x,t) \neq 0$ for all $(x,t) \in X \times J$ (hereafter we denote $\partial g/\partial t$ by $g'$).

2) $g(x,0) = g(x,1)$.

3) $g'(x,0) = g'(x,1)$.

We abbreviate "$g$ is a family of regular closed curves parametrized by $X$" by "$g \in FC^2(X)$".

If $g_0, g_1 \in FC^2(X)$, a <u>regular homotopy</u> between them, $G : g_0 \simeq g_1$ is a map $G \in FC^2(I \times X)$ for $I = [0,1]$ where $G(i,x,t) = g_i(x,t)$, $i = 0, 1$.

<u>Proposition 3</u>. <u>Let</u> $U = [0,\alpha) \subseteq J$. <u>Let</u> $X$ <u>be a connected manifold with boundary and let</u> $W \subseteq X$ <u>be a relatively open submanifold whose point set boundary</u>, $V$, <u>is a smooth submanifold</u>. <u>Let</u> $W_1$ <u>be a relatively open submanifold obtained from</u> $W$ <u>by deleting a collar of the form</u> $V \times [0,1]$.

*Let* $g_0, g_1 \in FC^2(X)$ *such that*

1) $g_0 = g_1$ *on* $(W \times J) \cup (X \times U)$.
2) $\mathrm{rot}(g_0 | x \times J) = \mathrm{rot}(g_1 | x \times J)$ *for any (hence all)* $x \in X$.
3) *If* $v_j = g_j'/|g_j'| : X \times J \longrightarrow S^1$, *then* $(v_0)_* = (v_1)_*$ *for* $(v_j)_* : H_1(X \times J) \longrightarrow H_1(S^1)$.

*Then there is a regular homotopy* $G : g_0 \simeq g_1$ *such that* $G_t \mid (W_1 \times J) \cup (X \times U)$ *is independent of* $t$.

This proposition obviously implies $\mathrm{Lemma}_1$ which as we've seen implies $\mathrm{Lemma}_n$, (Lemma 1").

Let $A = X - W_1$, $t_0 \in J - U$, and $g \in FC^2(X)$. For each $x \in A$, we straighten the curves $g(x,\cdot)$ in a neighborhood of $t_0$. Choose a smooth function $\varepsilon : X \longrightarrow [0,\infty)$ such that $\varepsilon \mid \overline{W}_1 = 0$, $\varepsilon \mid A > 0$ and such that for all $x \in X$, $[t_0-2\varepsilon(x), t_0+2\varepsilon(x)] \subseteq J - U$. By a regular homotopy it is easy to replace $g(x,t)$ by $g(x,t_0) + (t-t_0)g'(x,t_0)$ on $\bigcup_{x\in A} (x \times [t_0-\varepsilon(x), t_0+\varepsilon(x)])$ and leave $g$ unchanged outside $\bigcup_{x\in A} (x \times [t_0-2\varepsilon(x), t_0+2\varepsilon(x)])$. At every stage of the homotopy the map agrees with $g$ on $(W_1 \times J) \cup (X \times U)$. We will say that $g \in FC^2(X)$ is *straight along* $(A \times t_0)$ if such an $\varepsilon$ exists for which $g \mid x \times [t_0-\varepsilon(x), t_0+\varepsilon(x)]$ is linear in $t$ for all $x \in A$. The preceding remark is that any $g \in FC^2(X)$ can be changed by a regular homotopy that does not disturb $g$ on $(W_1 \times J) \cup (X \times U)$ so that it becomes straight along $A \times t_0$.

For any $g \in FC^2(X)$, define $\ell_g : X \longrightarrow (0,\infty)$ by

$$\ell_g(x) = \int_0^1 |g^1(x,t)|\,dt.$$

*Lemma 1. Let* $g \in FC^2(X)$, $U = [0,\alpha) \subsetneq J$, $W_1 \subseteq \overline{W}_1 \subseteq W \subseteq X$, $W_1, W$ *relatively open sets, and* $A = X - W_1$. *Suppose* $g$ *is straight along* $A \times t_0$ *for some* $t_0 \in J - U$. *Let* $m : X \longrightarrow [0,\infty)$ *be a smooth function such that:*

1) $m(x) \geqslant \ell_g(x)$ *for all* $x \in X$.
2) $(m-\ell_g) \mid \overline{W} = 0$.
3) *At each* $x_0 \in X$ *at which* $m(x_0) = \ell_g(x_0)$, $(m-\ell_g)$ *is flat (i.e. all partials vanish).*

*Then there is a regular homotopy* $G : g \sim \hat{g}$ *such that for all* $t$, $G_t = g$ *on* $(W \times J) \cup (X \times U)$ *and* $\ell_{\hat{g}} = m$.

*Proof.* Let $g(x,\cdot)$ be linear on $x \times [t_0-\varepsilon(x), t_0+\varepsilon(x)]$ for $x \in A$. Let $\vartheta : X \times J \longrightarrow [0,\infty)$ be a smooth function which is positive on $x \times [t_0-(\varepsilon(x)/2), t_0+(\varepsilon(x)/2)]$ for $x \in A$ and which vanishes on the complement of $\bigcup_{x\in A} \{x\times[t_0-\varepsilon(x),t_0+\varepsilon(x)]\}$. We assume that $[t_0-\varepsilon(x),t_0+\varepsilon(x)] \subseteq J - U$. Let $v : X \times J \longrightarrow S^1$ be such that $v(x,t) \perp g'(x,t)$ and that $g'(x,t) \wedge v(x,t) > 0$. (i.e. $\{g',v\}$ gives the usual orientation to $\mathbb{R}^2$.) Define $k : X \longrightarrow [0,\infty)$ by:

$$(1) \quad \int_0^1 \{|g'(x,r)|^2 + k^2(x)(\vartheta'(x,r))^2\}^{1/2} dr = m(x),$$

for all $x \in X$ and define the homotopy $G$ by

$$G_s(x,t) = g(x,t) + sk(x)\vartheta(x,t)v(x,t).$$

On the support of $\vartheta$, $g'(x,t) = g'(x,t_0)$ and so $v(x,t)$ is constant in $t$ there. Thus since $v \perp g$ we see that if $\hat{g} = G_1$, then

$$|\hat{g}'(x,t)| = \{|g'(x,t)|^2 + k^2(x)(\vartheta'(x,t))^2\}.$$

So (1) says that $\ell_{\hat{g}} = m$.

Since $k(x) = 0$ for $x \in \overline{W}$, $G_s(x,t) = g(x,t)$ for $(x,t) \in (\overline{W}\times J)$. Similarly the support of $\vartheta$ is disjoint from $X \times U$, so $G_s(x,t) = g(x,t)$ on $(X\times U)$ as well. We must merely check that $k$ is smooth. Except when $k$ vanishes, equation (1) can be used to give a formula for any derivative of $k$ in terms of the functions $|g'|^2$, $|\vartheta'|^2$, $m$ and $k$ itself. When $k$ vanishes we can use (1) to show that $k^2$ is flat whenever it vanishes. But the positive square root of such a function is obviously smooth. // (Lemma 1)

By virtue of Lemma 1 and the remarks preceding it, it suffices to prove Proposition 2 with the added hypothesis that $\ell_{g_0} = \ell_{g_1}$. This is obvious since the regular homotopy used to straighten $g_0$ and $g_1$ along $A \times t_0$ does not alter $g_0$ or $g_1$ in $(W_1\times J) \cup (X\times U)$ and in $W - W_1 \times J$ where $g_0 = g_1$, it changes them in the same way. Thus we may assume that the $g_0$ and $g_1$ of Proposition 2 are straight along $A \times t_0$. Apply Lemma 1 to $g_0$ and $g_1$ with $m : X \longrightarrow [0,\infty)$ satisfying $m(x) \geq \max(\ell_{g_0}(x),\ell_{g_1}(x))$. Since $W$ and $W_1$ are open submanifolds of $X$, whose boundaries in $X$ are smooth, any smooth function that vanishes on $W$ or $W_1$ is flat at each point of $\overline{W}$ or $\overline{W}_1$ respectively. Thus the flatness assumption on $m - \ell_{g_i}$ is automatic if we assume, $m - \ell_{g_i} > 0$ on $X - \overline{W}$ and vanishes on $\overline{W}$.

During the regular homotopies which stretch the curves $g_i(x,\cdot)$ so their lengths become $m(x)$, the maps $g_i$ are unchanged on

$(W\times J) \cup (X\times U)$. Of course, since $g_i$ is fixed on $X \times U$ during the course of both regular homotopies, we know that $\mathrm{rot}(g_i|x\times J)$ are unchanged throughout. Thus we prove Proposition 2 assuming $\ell_{g_0} = \ell_{g_1}$.

Remark. We also assume that for each $x \in X$, $g_0' \mid x \times U$ is not constant. This is a harmless assumption since if $g_0$ does not satisfy it, replace $g_0 = g_1$ on $X \times U$ with a map that does, giving us $\bar{g}_0$ and $\bar{g}_1 \in FC^2(X)$. The proposition, assuming $g_0' \mid x \times U$ is nonconstant for all $x \in X$, gives a regular homotopy $\bar{G} : \bar{g}_0 \simeq \bar{g}_1$ such that $\bar{G}_t \mid X \times U$ is independent of $t$. Thus replacing $\bar{G}_t \mid X \times U$ by $g_0 \mid X \times U = g_1 \mid X \times U$ gives us the required regular homotopy between $g_0$ and $g_1$.

Lemma 2. Given any $g \in FC^2(X)$, there is a diffeomorphism $(1\times\eta) : X \times J ↺$ such that if $h = g \circ (1\times\eta)$ then $h \in FC^2(X)$ and $|h'(x,t)| = \ell_g(x)$.

Proof. Let $L(x,t) = \int_0^t |g'(x,r)|dr$, and let $\ell(x) = \ell_g(x) = L(x,1)$. $L$ is smooth and $L'(x,t) = |g'(x,t)| > 0$. Thus $1 \times \lambda : X \times J \longrightarrow X \times J : x \longrightarrow (x,(L(x,t)/\ell(x)))$ is a diffeomorphism with inverse $1 \times \eta$. The chain rule gives

$$1 = \lambda'\circ(1\times\eta)\cdot\eta' = (L'\circ(1\times\eta)/\ell)\cdot\eta' = (|g'\circ(1\times\eta)|/\ell)\eta'.$$

Hence $|(g\circ(1\times\eta))'| = |g'\circ(1\times\eta)|\eta' = \ell$. // (Lemma 2)

Let $L_i(x,t) = \int_0^t |g_i'(x,r)|dr$ and $\ell_i = \ell_{g_i}$ and $\lambda_i = L_i/\ell_i$. For $x \in W$ since $L_0(x,t) = L_1(x,t)$ and $\ell_0(x) = \ell_1(x)$, we have $\lambda_0 \mid W \times J = \lambda_1 \mid W \times J$. This in turn implies that $\eta_0 \mid W \times J = \eta_1 \mid W \times J$. Similarly on $X \times U$, $(L_0-L_1)'(x,t) = |g_0'(x,t)| - |g_1'(x,t)| = 0$, so $L_0(x,t) - L_1(x,t) = r(x)$ on $X \times U$. But since $0 \in U$ and $L_0(x,0) = L_1(x,0)$ we see that $L_0 = L_1$ on $X \times U$ hence also on $X \times U$, $1 \times \lambda_0 = 1 \times \lambda_1 : X \times U \longrightarrow X \times \bar{U}$ where $\bar{U} = [0,\bar{\alpha})$ so $1 \times \eta_0 = 1 \times \eta_1 : X \times \bar{U} \longrightarrow X \times U$.

Now the map $(1\times\lambda_0)\circ(1\times\eta_1) : X \times J \longrightarrow X \times J$ is a diffeomorphism that is the identity on $(W\times J) \cup (X\times U)$. Consider the homotopy $H = 1 \times (s(\lambda_0(1\times\eta_1)) + (1-s)1) : X \times J \longrightarrow X \times J : (x,t) \longrightarrow (x,s(\lambda_0(x,\eta_1(x,t))) + (1-s)t)$. Since $\lambda_0(x,\cdot)$ and $\eta_1(x,\cdot)$ are increasing functions, this map is obviously 1:1 and the derivative of the second term with respect to $t$ is $s\lambda_0'(x,\eta_1(x,t))\cdot\eta_1'(x,t) + (1-s) > 0$ for all $s$, $t$. Hence $H_s$ is a diffeomorphism for each $s \in [0,1]$ which is the identity on $(W\times J) \cup (X\times U)$.

We remark that to prove Proposition 2 it suffices to prove it for $g_0\circ(1\times\eta_0)$ and $g_1\circ(1\times\eta_0)$. However since $g_1\circ(1\times\eta_0)\circ H$ is a regular homotopy fixed on $(W\times J)\cup(X\times U)$ which joins $g_1\circ(1\times\eta_0)$ with $g_1\circ(1\times\eta_1)$, it suffices to prove the proposition for $g_0\circ(1\times\eta_0)$ and $g_1\circ(1\times\eta_1)$. Thus we prove the proposition with the additional assumption that $|g_0'(x,t)| = |g_1'(x,t)| = \ell(x)$.

The map $g_j'/\ell : X\times J \longrightarrow S^1$ can be lifted to a map $\theta_j : \tilde{X}\times J \longrightarrow \mathbb{R}$ so that

$$g_j'(\pi(\tilde{x},t)) = \ell(x)e^{i\theta_j(\tilde{x},t)},$$

where $\pi : \tilde{X}\times J \longrightarrow X\times J$ is the universal covering space of $X\times J$. Write $v_j = g_j'/\ell$.

<u>Lemma 3</u>. $(v_0)_* = (v_1)_* : H_1(X\times J) \longrightarrow H_1(S^1)$ *implies that there is a smooth function* $\phi : X\times J \longrightarrow \mathbb{R}$ *such that*

$$\theta_1 - \theta_0 = \phi\circ\pi.$$

<u>Proof</u>. Let $\alpha : [0,1] \longrightarrow X\times J$ be any closed curve. Write $v_j\circ\alpha = e^{i\beta_j}$, where $\beta_0(0) = \beta_1(0)$. Since $(v_0)_* = (v_1)_*$, we know that $\beta_0(1) = \beta_1(1)$. Thus if $(v_1\cdot\bar{v}_0)\circ\alpha = e^{i(\beta_0-\beta_1)}$, then $(v_1\bar{v}_0)_* : H_1(X\times J) \longrightarrow H_1(S^1)$ is the zero homomorphism.

Suppose now that $\pi(\tilde{x},t) = \pi(\tilde{x}',t)$, we show that $(\theta_1-\theta_0)(\tilde{x},t) = (\theta_1-\theta_0)(\tilde{x}',t)$. Let $\tilde{\alpha}$ be any path in $\tilde{X}\times J$ connecting $(\tilde{x},t)$ with $(\tilde{x}',t)$. By definition of $\theta_j$, $v_1\bar{v}_0\circ(\pi\circ\tilde{\alpha}) = e^{i(\theta_1-\theta_0)\circ\tilde{\alpha}}$. Thus $(\theta_1-\theta_0)\tilde{\alpha}(1) = (\theta_1-\theta_0)(\tilde{\alpha}(0))$; $\theta_1-\theta_0$ is constant on the fibres of and thus defines a function $\phi : X\times J \longrightarrow \mathbb{R}$ satisfying $\phi\circ\pi = \theta_1-\theta_0$. // (Lemma 3)

Since $(W\times J)\cup(X\times U)$ is connected, we may assume that $\phi = 0$ on $(W\times J)\cup(X\times U)$.

We define the regular homotopy: Let

$$h_s(x,t) = g_0'(x,t)e^{is\phi(x,t)}$$

$$k_s(x,t) = h_s(x,t) - \int_0^1 h_s(x,r)\,dr$$

$$G_s(x,t) = g_0(x,0) + s(g_1(x,0) - g_0(x,0)) + \int_0^t k_s(x,r)\,dr.$$

Notice that for $j = 0, 1$, $h_j' = g_j'$; and since $\int_0^1 g_j'(x,r)dr = 0$ $k_j = g_j'$; and thus $G_j = g_j$, since $\int_0^1 k_j(x,r)dr = \int_0^t g_j'(x,r)dr = g_j(x,t) - g_j(x,0)$.

To see that $G_s$ defines a regular homotopy between $g_0$ and $g_1$:

1) $G_s(x,0) = G_s(x,1)$ since $\int_0^1 k_s(x,r)dr = 0$.

2) $G_s'(x,1) - G_s'(x,0) = k_s(x,1) - k_s(x,0) = h_s(x,1) - h_s(x,0) = g_0'(x,1)e^{is\phi(x,1)} = g_0'(x,0)e^{is\phi(x,0)} = 0$ since $g_0'(x,1) = g_0'(x,0)$ and $\phi(x,1) - \phi(x,0)$ is an integral multiple of $2\pi$.

As we already noted $\phi = 0$ on $(W\times J) \cup (X\times U)$ so for all $s$, $G_s = g_0$ on $(W\times J) \cup (X\times U)$.

Finally we must check that $G_s' = k_s$ is never zero. However, if $k_s(x,t) = 0$, then $v_0(x,t)e^{is\phi(x,t)} = \int_0^1 v_0(x,r)e^{is\phi(x,r)}dr$ or $1 = \int_0^1 (v_0(x,r)\bar{v}_0(x,t))e^{is(\phi(x,r)-\phi(x,t))}dr$. This implies that the integrand is identically 1; that is if for some $(s_0,x_0,t_0)$, $k_{s_0}(x_0,t_0) = 0$ then $h_{s_0}(x_0,t)$ is independent of $t$. However $h_{s_0} \mid x_0 \times U = g_0' \mid x_0 \times U$ and by our "harmless assumption," $g_0' \mid x_0 \times U$ is not constant so $k_s = G_s'$ never vanishes. // (Proposition 2 and Proposition 3 and the bijectivity $[I_\Sigma]_S \longrightarrow [I_S]_S$).

## Reader's Guide to Notation

Listed below are the notations used in this paper. They are arranged in order under the chapter and section in which they first appear in the text. Except where it seemed preferable to refer to the text, definitions or explanations of the terms are given.

### Chapter I

§1.1

$(X,Y)$ is the set of functions from $X$ to $Y$

$C(M,\mathbb{R}^2) = C^\infty(M,\mathbb{R}^2)$

$S = S(M,\mathbb{R}^2)$ is the set of smooth stable maps from $M$ to $\mathbb{R}^2$

$S(f)$ the singular set of $f$

$S_0 \subseteq S(f)$, the set of definite fold points

$S_1 \subseteq S(f)$, the set of indefinite fold points

$C \subseteq S(f)$, the set of cusp points

§1.2

$\pitchfork$, transverse

CPN of $f$ at $x$, coordinatized product neighborhood of $f$ at $x$

Transverse arc at $y \in \mathbb{R}^2$

Transverse manifold at $x \in M$, $T(x)$

§1.3

$W = W_f$, the space obtained from $M$ by identifying two points if they are in the same connected component of a fibre of $f$

$q = q_f$, the quotient map $q : M \longrightarrow W$

$\bar{f}$, the map that makes the diagram commute:

$$\begin{array}{ccc} M & \xrightarrow{f} & \mathbb{R}^2 \\ {\scriptstyle q}\downarrow & \nearrow_{\bar{f}} & \\ W & & \end{array}$$

$x \in S(f)$ is a simple singular point if $q^{-1}q(x) \cap S(f) = \{x\}$

$\Sigma = \Sigma_f = q(S(f))$

$\hat{\Sigma} = \hat{\Sigma}_f = q^{-1}(q(S(f))$

§1.4

$(\pm 1)$-cusp

conical neighborhood

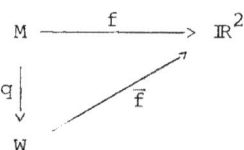

§1.5

$\mathcal{R}$, the set of components of $W - \Sigma$

$B(R) = q^{-1}(R)$, $R \in \mathcal{R}$

V, the set of q-images of cusp points and non-simple points of $S_1$

$\mathcal{C}$, the set of components of $\Sigma - V$

$\mathcal{C}_i$, the q-images of the components of $S_i - q^{-1}(V)$

$N(v)$, a conical neighborhood of $v$

$B(v) = q^{-1}(N(v))$

$N(c)$, a neighborhood in $W$ of $c \in \mathcal{C}$

$B(c) = q^{-1}(N(c))$

$T(c)$, a transverse manifold at any point on $q^{-1}(c) \cap S(f)$ for $c \in \mathcal{C}$

$T(v)$, a transverse manifold at $q^{-1}(v) \cap S(f)$ for $v \in V$

Y-, J-<u>bundles</u>

## <u>Chapter II</u>

§2.2

$\mathbb{A}_{(k)}$, a surface, $\mathbb{A}$, with k-discs removed

$C^2PN$, canonical coordinatized product neighborhood

## <u>Chapter III</u>

§3.0

$\mathcal{I}_f(M,\mathbb{R}^4) = \mathcal{I}$, the set of immersions of $M$ in $\mathbb{R}^4$ over $f$

$\mathcal{F}_f(M,\mathbb{R}^4) = \mathcal{F} = \{h \in C(M,\mathbb{R}^2) \mid (f,h) \in \mathcal{I}\}$

f-<u>regular</u> <u>homotopy</u>, a homotopy $(f,h_t)$ in $\mathcal{I}$

f-<u>fibre</u> <u>regular</u> <u>homotopy</u>, a homotopy $(h_t) \in \mathcal{F}$

$[\mathcal{I}]$, f-regular homotopy classes

$[\mathcal{F}]$, f-fibre regular homotopy classes

$B(\mathcal{R}) = \cup\{B(R) \mid R \in \mathcal{R}\}$

$r_h(R)$, rotation number of $h$ on $q^{-1}(x)$, $x \in R \in \mathcal{R}$

§3.1

$P_{\mathfrak{h}}$

§3.2

$n(R,c)$, the number of times the region $R \in \mathcal{R}$ abuts $c \in \mathcal{C}$

$Rot(\vartheta)$, the set of all functions $r \in \mathbb{Z}^{\mathcal{R}}$ such that $\sum_{R \in \mathcal{R}} r(R)n(R,c) = \vartheta(c)$, where $\vartheta$ is an orientation of $\mathcal{C}$

$Rot = \bigcup_{\vartheta} Rot(\vartheta)$

§3.3

$E = (f|S(f),e)$, a fixed embedding of $S(f)$ in $\mathbb{R}^4$

$E_i = (f|S(f),e_i)$, a fixed immersion of $S(f)$ in $\mathbb{R}^4$

$I_E = \{(f,h) \in I \mid (f,h) \mid S(f) = E\}$

$[I_E]$, the f-regular homotopy classes of elements of $I_E$

§3.4

$\delta_h$, the map from $C$ to $\{+1,-1\}$ such that $\delta_h(c) = +1\ (-1)$ if the germ of h on a transverse manifold to c is orientation preserving (reversing)

$\mathcal{D} \subseteq (C,\{\pm 1\})$

$v^c$, a chosen point on the component of $S(f)$ in the q-pre-image of a closed component $c \in C$

$v_c = q(v^c)$

$V_* = V \cup \{v_c \mid c,$ closed component of $C\}$

$V^* = q^{-1}(V_*) \cap S(f)$

$[I_A]_B$, the set of equivalence classes of elements in $I_A \subseteq I_B \subseteq I$ where $(f,h_0)$ and $(f,h_1)$ in $I_A$ are equivalent if there is an f-regular homotopy $(f,h_t)$ joining them and $(f,h_t) \in I_B$ for all t

$I_{V_*} \subseteq I_E$, the set of $(f,h) \in I_E$ whose germ at $V^*$ is fixed

$\hat{V} = q^{-1}(V) \cap S(f)$

$I_V \supseteq I_{V^*}$, the set of $(f,h) \in I_E$ whose germ at $\hat{V}$ is fixed

§3.5

$\tau_h : C \longrightarrow \mathbb{R}$, $\tau_h(c)$ is the total angle through which the image of the transverse manifold to c at $x \in c$ is turned by h as x moves along c, for $(f,h) \in I_V$

$\nu_h \mid C$, the integral part of $\tau_h$

$\omega_h \mid C = \omega_{\delta_h} \mid C$, the fractional part of $\tau_h$

$h_{\delta,\nu}$, the standard map germ at $S(f)$ determined by $(\delta,\nu)$ for $\delta \in \mathcal{D}$ and $\nu \in (C,\mathbb{Z})$

$H_S = \{$germs, $h_{\delta,\nu}$ at $S(f) \mid (\delta,\nu) \in \mathcal{D} \times (C,\mathbb{Z})\}$

$I_S = \{(f,h) \in I_{V_*} \mid$ germ of h at $S(f)$ is in $H_S\}$

§3.6

$H_A$ is a subset of the germs of maps at a set $\hat{A} \subseteq M$ into $\mathbb{R}^2$

$I_A = \{(f,h) \in I \mid$ the germ of h at $\hat{A}$ is in $H_A\}$

$H_{A,B}$ is the set of germs at $\hat{B}$ whose germs at $\hat{A} \subseteq \hat{B}$ are in $H_A$

$\hat{V}_* = V^* = q^{-1}(V_*) \cap S(f)$

$\hat{V} = q^{-1}(V) \cap S(f)$

$\hat{\Sigma} = q^{-1}q(S(f)) = q^{-1}(\Sigma)$

$\hat{S} = S(f)$

$A_\xi$, the set of oriented component arcs of $q^{-1}(\xi) - S(f)$ for $\xi \in \Sigma$

$A = \cup\{A_\xi \mid \xi \in \Sigma\}$

$\sigma_h : A \longrightarrow \mathbb{R}$, $\sigma_h(a)$ is the rotation number of $h$ on the arc, $a \in A$ for $h \in H_{S,\Sigma}$

$\nu_h \mid A$, the integral part of $\sigma_h$

$\omega_h \mid A = \omega_{\delta_h} \mid A$, the fractional part of $\sigma_h$

$\lambda : H_{S,\Sigma} \longrightarrow \mathcal{D} \times (C \cup A, \mathbb{Z}) : h \longrightarrow (\delta_h, \nu_h) = \lambda_h$

$L_\Sigma = \lambda(H_{S,\Sigma})$

$H_\Sigma$ a subset of $H_{S,\Sigma}$ making $\lambda$ a 1:1 correspondence onto $L_\Sigma$

$I_\Sigma = \{(f,h) \in I_S \mid (h)_{\hat{\Sigma}} \in H_\Sigma\}$

§3.7

$A^* = \cup\{A_\xi \mid \xi \in V_*\}$

<u>arc-word</u>

<u>cycle word</u>

$N_\delta$, for $\delta \in \mathcal{D}$, $N_\delta = \{\nu \in (C \cup A^*, \mathbb{Z}) \mid (\delta,\nu) = (\delta_h,\nu_h)$ for some $(h) \in H_{S,\Sigma}\}$

$\vartheta_\delta(c) = (-1)^i\delta(c)$ for $\delta \in \mathcal{D}$, $c \in C_i$

$N_\delta^! = \{\nu \in N_\delta \mid (\delta,\nu)$ determine, uniquely, an element $\tilde{r}(\delta,\nu) \in \mathrm{Rot}(\vartheta_\delta)\}$

$L_\Sigma^! = \cup\{\delta \times N_\delta^!\}$

§3.8

$N_\delta^* = \{\nu \in N_\delta^! \mid (\delta,\nu) = (\delta_h,\nu_h)$ for some $(f,h) \in I_S\}$

$L = \cup\{\delta \times N_\delta^* \mid \delta \in \mathcal{D}\} = \lambda(I_S)$

$m(R)$, the number of handles of $R$

$n(R)$, the number of boundary components of $R$

$Z = \underset{R \in \mathcal{R}}{\times} (\mathbb{Z}^{2m(R)})$

$\zeta : [I_S]_S = [I_\Sigma]_\Sigma \longrightarrow Z$

$P = L \times Z$

$\pi = (\lambda,\zeta) : [I_\Sigma]_\Sigma \longrightarrow L \times Z$

§3.9

$G = \underset{R \in \mathcal{R}}{\times} (\mathbb{Z}^{n(R)-1})$

$\gamma : I_\Sigma \longrightarrow G$

$Q = L \times Z \times G$

§3.10

$P : [I_\Sigma]_\Sigma \longrightarrow [I]$, the obvious projection

$sp_H : \hat{V} \longrightarrow \mathbb{Z}$ for $H$ an f-fibre regular homotopy

§3.11

$\Sigma^* = q^{-1}(V_*) \cup S(f)$

$N_\delta$, the image of $N_\delta^*$ in $H^1(\Sigma^*)$

## Index

# References


[B] S. J. Blank. Extending immersions of the circle. Thesis, Brandeis University (1967). (See Poenaru Exposé 342, Sem. Bourbaki (1967-68) Benjamin, N. Y.)

[B-C] S. J. Blank and C. Curley. Desingularizing maps of corank one. PAMS vol. 80, number 3 (1980) 483-486.

[B-deR] Oscar Burlet and Georges deRham. Sur certaines applications génériques d'une variété close a 3 dimensions dans le plan, L'Enseignment Math. 20 (1974) 275-292.

[E] C. Ehresmann. Sur les espaces fibrés differentiables. C. R. v. 224 (1947) 1611-1612.

[E-G] Ja. M. Eliasberg and M. L. Gromov. Removal of singularities of smooth mappings. Izv. Akad. Nauk SSR series Mat 35 (1971) 600-626 = Math. USSR Izv. 5 (1971) 615-639.

[F-T] G. K. Francis and S. Troyer. Excellent maps with given folds and cusps. Houston J. of Math. vol. 3, No. 2 (1977) 165-194. Continuation in Houston J. of Math., vol. 18, 53-59.

[$G^2$] M. Golubitsky and V. Guillemin. Stable mappings and their singularities. Grad. Texts in Math. 14 Springer 1973.

[G-W] T. Gaffney and L. Wilson. Equivalence of generic mappings and $C^\infty$ normalizations. Compositio Mathematica (to appear).

[G-du P-W] T. Gaffney, L. Wilson and A. du Plessis, Map germs determined by their discriminant (in preparation).

[H] M. Hirsch. Differential Topology. Grad. Texts in Math. 33 Springer (1976).

[Haef] A. Haefleger. Quelques remarques sur les applications d'une surface dans le plan. Ann. Inst. Fourier (Grenoble) 10 (1960) 47-60.

[Hart] Robin Hartshorne. Algebraic Geometry. Grad. Texts in Math. 52 Springer (1977).

[K] L. Kushner. On maps from 3-manifolds to the plane, Thesis, Brandeis (1980).

[K-L-P] L. Kushner, H. Levine, and P. Porto. Mapping three manifolds into the plane. Bol. Soc. Mat. Mex. vol. 29 no. 1 (1984) 11-34.

[Lang] S. Lang. Introduction to Differentiable Manifolds. Interscience (1962).

[$L_1$] H. Levine. Singularities of differentiable mappings. Proc. of Liverpool Singularities I. Lecture Notes in Math. 192 Springer (1971) 1-89.

[$L_2$] H. Levine. Elimination of cusps. Topology 3 (1965) 263-296.

[Lima] E. Lima. On the local triviality of the restriction maps for embeddings. Comm. Math. Helv. 38 (1964) 163-164.

[Mather] J. Mather. Stability of $C^\infty$-mappings V. Advances in Mathematics, vol. 4, no. 3 (1970) 301-336.

[Mo] B. Morin. Formes canonique des singularites d'une application différentiable. C. R. 260 (1965) 5662-5665, 6503-6506.

[P] R. S. Palais. Local triviality of the restriction map for embeddings. Comm. Math. Helv. 34 (1960) 305-312.

[T] R. Thom. Ensembles et morphismes stratifies. Bull. Amer. Math. Soc. vol. 75, no 2 (1969) 240-284.

[W] H. Whitney. On regular closed curves in the plane. Compositio Math. v. 4 (1937) 276-284.

- Biogas energy was exempted from the $CO_2$ tax in Sweden, and biofuel vehicles was benefited from a subsidy of Euro 1,100 (Puertas 2012): complementary product.
- Electricity and heat were relatively cheap in Sweden.
- Gasoline was priced at about Euro 1.506 per liter (SEK 13.40) and diesel at about Euro 1.494 (SEK 13.29) and daily consumption accounted for over 5 % of average daily wages.[3]

By mid-March 2014, Anna-Karin Hatt, the Swedish Minister of Information Technology and Energy, inaugurated the GoBiGas demonstration plant for large-scale production of biogas. The project by *Göteborg Energi* was heralded as one that took biogas production to an entirely new level. Its success was to decisively alter the dynamics of sustainable transportation in the country and was claimed to be the first step to a fossil-fuel-free future. "Stage 2 of GoBiGas is open for cooperation with new interested parties. The project has been allocated approximately €58.8 million (app SEK 520 million) in support from the EU, provided the technology and performance can be verified in stage 1, and that the project can be financed and judged to be profitable."[4]

---

[3] http://www.bloomberg.com/slideshow/2013-02-13/highest-cheapest-gas-prices-by-country.html#slide8.

[4] http://www.businessregiongoteborg.com/newsarchives/newsarticles2014/inaugurationofthegobigasbiogasplant.5.5783eddf144e09ec4fef3f33.html, updated 27 March 2014, last accessed 27 April 2014.

ANNEX: *Göteborg Energi* Financial Performance 2003–2012

| | 2012 | 2011 | 2010 | 2009 | 2008 | 2007 | 2006 | 2005 | 2004 | 2003 |
|---|---|---|---|---|---|---|---|---|---|---|
| Net sales | 6,956 | 7,415 | 7,791 | 5,747 | 4,009 | 3,687 | 3,567 | 3,324 | 3,232 | 3,147 |
| Raw materials and supplies | −3,997 | −4,375 | −4,623 | −3,170 | −1,708 | −1,486 | −1,521 | −1,291 | −1,233 | −1,581 |
| Depreciation/amortization | −773 | −772 | −730 | −619 | −584 | −536 | −463 | −486 | −509 | −402 |
| Operating profit | 742 | 658 | 930 | 691 | 709 | 719 | 714 | 685 | 664 | 437 |
| Net financial income/expenses | −227 | −204 | −115 | −70 | −198 | −155 | −88 | −71 | −92 | −78 |
| Profit after financial items | 515 | 454 | 815 | 604 | 510 | 565 | 626 | 615 | 572 | 359 |
| Fixed assets | 12,443 | 11,814 | 11,216 | 10,704 | 10,165 | 9,316 | 8,421 | 7,574 | 6,436 | 5,759 |
| Equity | 5,087 | 4,680 | 4,561 | 4,062 | 3,785 | 3,461 | 3,308 | 3,013 | 2,759 | 2,586 |
| Non-current liabilities | 4,833 | 4,304 | 4,231 | 4,696 | 4,811 | 4,127 | 3,492 | 2,908 | 2,461 | 1,945 |
| Return on equity (%) | 7.6 | 7 | 13.5 | 11 | 9.8 | 11.6 | 13.8 | 14.8 | 14.9 | 10.3 |
| Return on total capital (%) | 5.2 | 4.8 | 7 | 5.7 | 6.5 | 7.1 | 7.8 | 8.6 | 9.5 | 6.9 |
| Return on capital employed (%) | 5.8 | 5.4 | 8 | 6.5 | 7.6 | 8 | 8.7 | 9.7 | 10.8 | 7.9 |
| Equity/assets ratio (%) | 34.8 | 33.7 | 34.1 | 32.2 | 33.5 | 33.4 | 34.6 | 35.5 | 36.8 | 38.1 |
| Investment | 1,470 | 1,371 | 1,274 | 1,253 | 1,326 | 1,404 | 1,354 | 1,628 | 1,174 | 689 |
| Production of electricity incl. CHP (GWh) | 668 | 973 | 1,177 | 1,082 | 637 | 902 | 340 | 144 | 176 | 419 |
| Electricity supply network (electricity transmission) (GWh) | 4,720 | 4,849 | 4,999 | 4,786 | 4,858 | 4,859 | 4,951 | 4,902 | 5,077 | 5,009 |
| Sales of district heating (GWh) | 3,876 | 3,647 | 4,470 | 3,864 | 3,508 | 3,558 | 3,644 | 3,737 | 3,843 | 3,876 |
| Sales of natural gas and biogas (GWh) | 945 | 663 | 762 | 643 | 687 | 651 | 690 | 681 | 792 | 727 |
| Average number of employees | 1,173 | 1,189 | 1,164 | 1,108 | 1,018 | 981 | 1,006 | 998 | 992 | 1,092 |
| Emissions of fossil carbon dioxide (kt) | 314 | 480 | 649 | 545 | 301 | 455 | 290 | 136 | 175 | 265 |

Amounts in SEK million

# Teaching Note

## *Case Synopsis*

In the year 2008, Business Region Goteborg, a collection of 13-member municipalities, was awarded the "Blue Sky Innovation Award" for encouraging the use of biomethane as a sustainable transportation fuel. Goteborg, in particular, and Sweden, in general, had viewed the increased use of biogas as key to creating a more sustainable society and as a step toward a fossil-fuel-free future. Goteborg Energi's Euro 150 million/20 MW, growing to 100 MW Gothenburg Biomass Gasification Project ("GoBiGas") commissioned in late 2013 and formally inaugurated in March 2014, was designed to gasify 50,000 t of forest wastes and wood pellets per year and to produce gas to fuel 80,000–100,000 cars. This first of its kind plant generated gas that was similar to natural gas. The project brought together gasification and methanation technologies that were proven separately. The European Union had offered grants for expansion subject to the success and economic viability of the 20 MW pilot project.

## *Case Question*

Electricity being relatively lower priced in Sweden, the project sought to displace more expensive transportation fuel. The case seeks to lead to the development of alternative scenarios for evolution in pricing of transportation fuels that would enhance the viability of the pilot project and pave the way for its expansion.

## *Teaching Objectives*

- Enabling participants to appreciate the technology risk mitigation measures put in place by the project.
- Highlight the positioning of biogas as a transportation fuel, rather than using it for electricity generation.
- Enable participants to alter input values and assess the sensitivity of project viability to key variables.
- For participants to compare alternative transportation fuel pricing scenarios and assess the displacement possibilities of biogas generated by the project.

## *Case Objectives and Use*

The biogas from the project is slated to displace petroleum derivatives in transportation. The case deals with the analysis of relative pricing of fuels, which, in turn, are a function of input costs of wood and residues for the GoBiGas Project and of international crude oil prices for the refineries. Additionally, prevailing electricity tariffs were quite low, and the biogas from the plant would not have been competitive relative to such tariffs. Sweden had enunciated a goal of achieving a fossil-free fleet by the year 2030. GoBiGas—phases 1 and 2—would go a long way toward meeting this target.

However, in the medium term, there could be a large-scale switch to hybrid vehicles or to electric vehicles, and at that stage, GoBiGas might need to compete against low electricity tariffs. The discussion could be structured to project electricity tariffs and petroleum distillate prices and to compare the prices with biogas delivered by the project.

The present project is largely grant-funded, and success of the experimental setup would be measured by merely being able to prove the concept. Subsequent plants might be built at lower costs.

The instructor could choose to expand the scope of the case to include other aspects that could impact the decision such as being able to place a price on the environmental externality and being able to feed the gas into an existing generator and generate electricity.

## *Teaching Plan*

The project discussed herein is a stand-alone initiative, and the first of its kind implemented anywhere in the world, and hence, has received worldwide attention and state funding. Yet, the project promoters, *Goteborg Energi,* have managed to minimize technology risk exposure by employing proven technology modules and experts. The risks faced by the project are more operational and financial than technical: ensuring consistent and continued supply of feedstock, efficient plant operations, competing against fossil fuel prices, etc. The analysis for Phase 1 could include a comparison of landed costs of biogas and market prices for transportation fuels. The threshold price of the nearest plausible substitute would determine viability of the biogas project.

The second phase could be analyzed with both transportation fuels as well as electricity (for hybrid vehicles as well as electric vehicles) as potential competitors, and the analysts could determine the threshold values that would render biogas competitive relative to the substitutes.

The instructor could guide the students to evaluate breakeven value in a number of small steps:

1. **Levelized cost of energy (LCOE) for biogas**: recasting the fixed and variable costs into an annuity framework and computing the cost of generating 1 MW equivalent.
2. **Projected pricing for petrol/diesel**: Data available online could be collected from one of many sources/agencies. This would then need to be adjusted to reflect the tax structures, etc., for fuels in Sweden/Göteborg.
3. **Projected pricing for electricity**: Data available online could be collected from one of many sources/agencies. This would then need to be adjusted to reflect the tax structures, etc., for fuels in Sweden/Göteborg.
4. **Graphical analysis of the scenarios**: Analysts could plot data to generate various convergence scenarios to determine a range of prices at which biogas from the project would be cost-competitive relative to its nearest substitute(s).

## References


GoBiGas (2014) GoBiGas meets the growing need for biogas. http://gobigas.goteborgenergi.se/En/About_us. Accessed 1 Apr 2014

Gunnarsson I (2014) Efficient transfer of biomass to Bio-SNG of high quality. Nordic Baltic Bioenergy 2013, p 19 of 20. http://nobio.no/upload_dir/pics/Ingemar-Gunnarsson.pdf. Accessed 2 Apr 2014

Hannula I, Kurkela E (2014) Low grade fuel to high quality energy by gasification. Technical Research Center of Finland VTT, International Conference on Thermochemical Conversion Science, 11 Sept 2013. http://www.gastechnology.org/tcbiomass2013/tcb2013/02-Hannula-tcbiomass2013-presentation-Wed.pdf. Accessed 28 Mar 2014

Lewald A (2014) Swedish bioenergy projects/strategies. Swedish Energy Agency. http://lnu.se/polopoly_fs/1.35420!Swedish%20bioenergy.pdf. Accessed 28 Mar 2014

Messenger B (2014) Valmet supplies first for kind gasification plant producing biofuel for transport in Sweden. Waste Management World, 14 Mar 2014. http://www.waste-management-world.com/articles/2014/03/valmet-supplies-first-for-kind-gasification-plant-producing-biofuel-for-transport-in-sweden.html?cmpid=EnlWMW_WeeklyMarch142014. Accessed 28 Mar 2014

NETL (2014) National Energy Technology Laboratory, United States Department of Energy. http://www.netl.doe.gov/File%20Library/Research/Coal/energy%20systems/gasification/worldwide%20database/GasificationDB2010.xlsx. Accessed 1 Apr 2014

Puertas J (2012) Renewable gas: the sustainable energy solution. Program Committee on Sustainability, International Gas Union, 2009–2012 Triennium Work Report, Kuala Lumpur World Gas Conference, June 2012, p 31 of 52